Ethical Issues in Counselling and Psychotherapy Practice

Poornima Bhola · Ahalya Raguram
Editors

Ethical Issues in Counselling and Psychotherapy Practice

Walking the Line

Springer

Editors
Poornima Bhola
Department of Clinical Psychology
National Institute of Mental Health
and Neurosciences (NIMHANS)
Bengaluru, Karnataka
India

Ahalya Raguram
Department of Clinical Psychology
National Institute of Mental Health
and Neurosciences (NIMHANS)
Bengaluru, Karnataka
India

ISBN 978-981-10-1806-0 ISBN 978-981-10-1808-4 (eBook)
DOI 10.1007/978-981-10-1808-4

Library of Congress Control Number: 2016942890

Printed on acid-free paper

This Springer imprint is published by Springer Nature
The registered company is Springer Science+Business Media Singapore Pte Ltd.

Foreword

This book, *Ethical Issues in Counselling and Psychotherapy Practice: Walking the Line*, is an experiential narrative of varied clinical dilemmas concerning ethical conduct inherent in mental health service delivery in different settings across India. The code of professional ethical conduct is a logical derivative of a set of human values that are common across races, cultures and religions. Human values are meant to ensure individual and collective wellbeing of mankind. The other derivative is the system of justice whose basic tenets are again similar across societies. The code of conduct prescribed for mental health professionals is a small part of this system of justice. The clinical-therapeutic dilemmas occur when any of the various components of the clinical-therapeutic process in a given situation happens to or appears to violate the purpose of a code of conduct.

A brief overview of the role of human values in mental health service delivery is warranted in order to appreciate the contents as well as the usefulness of this book to both mental health professionals and common public. For this purpose, I prefer to use the ancient Indian tradition as an example, as I am more familiar with it and I believe that parallel templates are available in scriptures of other religions also.

The system of values in the Indian tradition is called 'dharma'. It has many hierarchical meanings depending upon the level of application. When applied to social interaction, dharma manifests as the principle of 'righteousness', consisting of a set of values, attitudes and behaviour that is conducive to the greater good of all. The basis of this principle is the fact of the original 'Universal-One' becoming the 'Manifest-Many' through the process of Creation, irrespective of one's belief system, viz. 'ancient-religious' or 'modern scientific'. Obviously, the responsibility of upholding this principle of righteousness rests upon the most evolved of all life forms, mankind. Thus, the adoption of 'righteous-living' in day-to-day life is a mandatory goal for both the individual and society to aspire to achieve at all levels and aspects of living. This goal has four component objectives.

The first objective is the 'realization' by direct experience of the 'Universal-One-ness' in one's life, amidst the Manifest-Many. It is also known as 'God-realization', and the topic lies beyond the scope of this foreword.

The second is the pursuit of an ideal of positively contributing to the society's wellbeing which in turn ensures one's own wellbeing. Ayurveda, the ancient Indian system of wellbeing, prescribes the following attitudes and behaviour to fulfil this objective: (i) Cultivate affection, forgiveness, helpfulness and absence of jealousy towards others. (ii) Besides living in harmony with the environment and other human beings, foster harmony between one's needs, abilities and limitations. (iii) Whenever an occasion arises, help a fellow human in distress by comforting, consoling, instilling hope and confidence and helping to manage issues caused the distress, to the best of one's ability. Such behaviour is considered as the 'highest human virtue'. Interestingly, this 'helping' sounds very similar to what we now call psychotherapy/counselling.

The third objective is to be fair (just) in all social interactions reflected by 'doing unto others what one would like others to do unto oneself'. This maxim demands a readiness to sacrifice certain seemingly personal advantages whenever such an advantage involves potential difficulties to others. Every mental health professional is acutely aware that the violation of this maxim is the root cause of all interpersonal conflicts.

The fourth is the application of components of righteousness to the many roles in an individual's routine life, such as the duties and the responsibilities as a king, minister, soldier, head of the family, husband, wife, teacher or healer. For example, a physician is expected to cultivate the qualities of sincerity, righteousness and purity of mind (freedom from the 'pettiness of the small-self'), compassion and wisdom (expertise in decision making). The modern equivalent of this description of the duties of a healer is the set of ethical principles approved by a recognized authority that guides the delivery of mental health services.

The foregoing description triggers a legitimate question: If the act of helping a fellow human in distress is the highest human virtue as a natural component of human values, how can or how does such a process become a source of ethical dilemmas? There are three interacting causes for this. They are consequences of the technology-dominant money-oriented value-system of the modern human culture. Even health services have become organized industries of differing sizes and complexities at state, corporate or individual levels.

(a) The first cause is the ever-increasing conflict between the 'I' and 'thou', the latter representing both 'you' and 'them'. The conflict concerns the diffuse boundary between individual autonomy and one's responsibility towards collective wellbeing. The heavily weighted sense of individual autonomy in the modern times contributes to most dilemmas, particularly in circumstances which demand sacrifice of some personal agenda.
(b) The second is the direct consequence of 'modernity'. In ancient times, helping a fellow human in distress was an unselfish and sincere act of free choice without expectation of any reward, in an atmosphere of mutual trust and confidence. It often involved a certain degree of sacrifice on the part of the helper. In contrast, currently, 'helping' has become an organized trade as 'health services' with polarization of the service-provider and the customer.

The former is burdened with certain responsibilities and the latter awarded certain rights. Even in a setting of free service, somebody will be paying for the expenses and labour. Most of the time, the overall atmosphere is one of 'maximum gain with least investment/effort.'

Closely related to this 'modern industrial culture' are two mindsets that burden the mental health service-setting. One is the effect of media advertisements over many generations, 'brainwashing' the modern service-receiver to expect and even demand 'easy solutions' or an 'easy route' to wellbeing! The other is the existential fact of the human tendency to deny the unpredictable variability of potentially stressful life-events and attempt to control or manipulate them. This attitude expects and even demands precise predictions of outcomes of treatment. This attitude is analogous to expecting assurances of successful outcomes with no side effects associated with treatments.

(c) The third cause is the universal and historic fact of the natural, inherent gap between the ideal and the actual in all areas of human endeavour. The history of human culture is a narrative of this perpetual striving to realize or actualize the ideals despite falling short of them all the time. In respect of ethics in mental health services, this 'ideal–actuals' gap manifests as a problem of interpreting and translating ethical principles into professional behaviour in real-life clinical situations. The ideal principles are necessarily rendered in the form of clearly defined language for wider discourse. The 'ideal principles' constitute the 'spirit of the law', and the language format constitutes the 'letter of the law'. The problem of transforming the word meanings ('letter of the law') into decisions and actions so as to satisfy the ideals ('spirit of the law') is a major cause of dilemmas. This is similar to the problem when a court of law assesses the correspondence between decisions and actions of a mental health professional and the intended goals. But, the eternal ideal–actual gap should never be an excuse to give up the ongoing effort or struggle to behave ethically in helping relationships.

This description traces the numerous ways in which what was a simple process of human kindness as an ideal to help a fellow human in distress has become a potential source of ethical dilemmas in the present-day commercial atmosphere. Adding insult to injury, these dilemmas are also capable of getting precipitated into damaging legal proceedings. It is unlikely that any mental health professional will encounter all the possible dilemmas. Yet, in accordance with the principle of 'prevention is better than cure', it is imperative that professionals are familiar with the range and nature of potential dilemmas.

It is in this context that this book offers readers a wide range of ethical dilemmas from an equally wide range of real-life clinical situations, including circumstances when the social matrix itself is not entirely healthy. These vivid and authentic accounts are rendered in the form of experiential narratives enabling readers to empathize easily and reflect upon the intricacies of the conflicts involved. I believe that these reflections contribute not only to preventive-alertness and caution about potential ethical dilemmas but also to what can be called 'living-learning'. For these reasons, I believe that this book satisfies the needs of a mental health professional in the field of ethics and hope that it will eventually qualify as a 'must-read'.

In all varieties of settings of mental health service delivery, active and cooperative participation of the client/patient and caregivers is absolutely essential. For this very reason, it is essential for the general public as potential users of mental health services to become aware of the kind of ethical dilemmas that can vitiate the therapist–client and professional–caregiver relationships.

In this regard, I believe this book to be eminently qualified for educating the interested general public about ethical issues concerning the interface between the mental health services and its users.

Dr. C. Shamasundar
Retired Professor of Psychiatry
NIMHANS, Bengaluru

Acknowledgements

This book represents the culmination of ideas, experiences and critical questions about the training and practice of psychotherapy and counselling. As teachers and trainers of psychotherapy, the editors of this book were struck by the inadequate emphasis given to ethical issues in most psychotherapy and counselling training programmes. Even more glaring was the absence of any reliable and easily accessible resource to guide practitioners as they attempted to 'walk the line' in their work with clients in diverse practice settings. This lacuna provided the impetus to organize a symposium on 'Ethical issues in Psychotherapy Practice' on 26 and 27 November 2013 at the National Institute of Mental Health and Neuro Sciences, Bengaluru, India. The symposium brought together seasoned practitioners as well as new entrants in the field of psychotherapy, to share experiences, dilemmas, mistakes and reflections in the common endeavour of providing high-quality and ethically sound therapeutic services. What resulted was a rich tapestry that wove together the experiences of practitioners working in very diverse settings, with a range of populations who encountered complex challenges posed by emerging ethical issues. A dialogue on these issues was a strongly felt need in the professional community and was also perhaps long overdue. The idea of bringing together the rich insights shared and gained during the days of the symposium seemed to be the natural outcome of this. We owe a debt of gratitude to many who have contributed to the fruition of what began as an idea.

We gratefully acknowledge the generous grant from the Dr. Ramachandra N. Moorthy Foundation, which enabled us to invite eminent speakers and practitioners from all over the country for the symposium. This gave a truly representative flavour of the diversity of issues and populations and also illuminated the many common challenges and concerns.

Our deepest gratitude to all the contributing authors, many of whom were speakers at the symposium, for sharing their experience and wisdom. The honesty and openness in sharing their reflections has been striking and we are grateful for their generous investment of time in putting their ideas together in the written form for the book. We thank our contributors for their long hours of patient and thoughtful work in response to requests for revisions of their chapter drafts.

The enthusiastic participants of the symposium deserve our thanks as well. Their active participation, vigorous debate and questioning fed back into the chapters and greatly enriched all of them.

We deeply value the support and encouragement of our former Director/Vice Chancellor, Prof. P. Satish Chandra and former Registrar, Prof. V. Ravi which catalyzed the initiative to bring out a comprehensive publication. We also acknowledge the ongoing support of our current Director/Vice Chancellor, Prof. B.N. Gangadhar and Registrar, Prof. K. Sekar.

Our colleagues at the Department of Clinical Psychology, who have directly and indirectly helped us in numerous ways, deserve our grateful thanks. Their generosity with their ideas and their time, both during the organization of the symposium and in many ways thereafter, have been invaluable in completing this task.

Our young colleagues in the department, Ms. Manila Mathews and Ms. Ananya Sinha, deserve a special note of appreciation for gamely shouldering the numerous demands we have made of them. They have spent many hours on checking references, formatting and in doing many other tasks that required meticulous attention to detail. Our thanks to Ms. Madhuri Dugyala for intensive background work for one of the chapters.

We thank Dr. C. Shamasundar, former Professor of Psychiatry, NIMHANS, for sharing his wisdom and writing the foreword for this book. We are also very grateful to Ms. Shinjini Chatterjee, Ms. Shruti Raj Srivastava and the excellent editorial team at Springer for providing a platform for this book and for their support through the process.

Putting this book together has been an enriching learning experience for us. We hope that readers will find it stimulating and useful for their professional growth and practice. The objectives of the book will be served if the content validates their experiences, challenges existing beliefs and encourages readers to reflect from a different vantage point. Most importantly, we hope there are more spaces and contexts created for learning, teaching and thinking about the ethics of practice.

Poornima Bhola
Ahalya Raguram

Contents

Editors and Contributors

Editors

Poornima Bhola, Ph.D., is Associate Professor in the Department of Clinical Psychology, National Institute of Mental Health and Neuro Sciences (NIMHANS), Bengaluru, India. She has over 18 years of experience as a therapist and has previously worked in a consultation-liaison context at St. John's Medical College & Hospital, Bengaluru. Her keen interest in ethical questions is reflected in her involvement in teaching, training and research concerning the ethics of practice. She is a member of the Ethics Committee of the Indian Association of Clinical Psychologists and of the Institutional Ethics Committee at Karunashraya Trust. She is deeply interested in understanding and enhancing the professional training and development of psychotherapists and has also been involved in the *International Study of the Development of Psychotherapists*. In addition, her clinical and research interests include psychotherapy processes, youth mental health and suicidality, personality disorders, psychosocial rehabilitation and workplace mental health.

Ahalya Raguram, Ph.D., is Professor at the Department of Clinical Psychology and the former head of the department, NIMHANS, Bengaluru, with over 25 years of experience. Her clinical, teaching and research interests reflect a long-standing and abiding interest in families and interpersonal relationships as well as individual, family and couple therapies. Her research work has focused on exploring the role of client and therapist factors in the outcome of therapies in addition to examining the efficacy of different approaches to intervention. She is the coordinator of the psychotherapy training programme in the department. In addition she has conducted numerous training programmes to impart basic counselling and therapeutic skills to various groups including counsellors from family counselling and rehabilitation centres, lay counsellors, nursing staff as well as school and college teachers. She was an expert member of the Advisory Committee of the Indira Gandhi National Open University, New Delhi, for curriculum development of a course in family counselling. She has been a member of the Institutional Ethical Review Board at NIMHANS.

Contributors

Chandana Baksi is a certified psychotherapist and trainer, with a special emphasis on gender and mental health. She has been working with Samikshani (a Kolkata-based mental health organization) and Swayam (a Kolkata-based women's rights organization) for the last two decades. She has written a number of articles and training manuals as well as conducted numerous trainings in India, Bangladesh, Nepal and Pakistan on a wide range of mental health and gender related issues. She also served as faculty of the PGDMH (Post Graduate Diploma in Mental Health) Course at Jadavpur University, Kolkata.

Sowmya Bhaskaran T.S. is a graduate of Mahatma Gandhi Medical College, Mumbai and a postgraduate in psychiatry from Madras Medical College, Chennai. She recently completed her Doctorate in Medicine in child and adolescent psychiatry at NIMHANS, Bengaluru. At present, she works as a consultant child and adolescent psychiatrist at Insight Clinic in Coimbatore, India. Her areas of interests include working with survivors of child sexual abuse and children with developmental disorders.

Shruti Chakravarty is a mental health professional, trainer, researcher and activist, who has been engaging with issues of gender and sexuality for the last 12 years. Currently, Shruti has an independent counselling practice based out of Mumbai and is pursuing her Ph.D. at the Tata Institute of Social Sciences, Mumbai. She has previously worked as a counsellor on a telephone helpline for lesbian women from 2004 to 2007. She has been involved in two research studies on violence against disabled, lesbian and sex-working women in Bangladesh, India, and Nepal and on conceptualizing gay affirmative counselling practice in India. Additionally, Shruti has presented papers related to her areas of interest in both national and international conferences.

Buli Nag Daimari is a psychiatric social worker in the Department of Psychiatric Social Work, LGB Regional Institute of Mental Health (LGBRIMH), Tezpur, Assam. She has an M.Phil. in Psychiatric Social Work from NIMHANS and is currently pursuing a Ph.D. at the Tata Institute of Social Sciences, Mumbai. She is associated with the Family Therapy Centre at LGBRIMH, Tezpur. She has also completed her diploma in special education for mental retardation from the Composite Regional Centre, Guwahati Medical College Hospital, Guwahati, affiliated to the National Institute of Mentally Handicapped, Secunderabad. Her research interests are in the area of family studies, marriage and mental health, children and adolescent mental health, addiction, and disability. She is also a paralegal volunteer, District Legal Services Authority, Sonitpur, Assam.

Neena David is a clinical psychologist with over 18 years of experience in working with children, adolescents, teachers and school communities. Currently, in addition to her independent practice, she provides counselling services at the Mallya Aditi International School, Bengaluru. She has been instrumental in setting up processes and consultation services for school mental health programmes across a range of schools. Neena has a Masters in Applied Psychology from Women's Christian College, Chennai, an MPhil in Clinical Psychology from NIMHANS, and a doctorate from the Centre for Human Ecology, Tata Institute of Social Sciences, Mumbai. Her research interests include school mental health, inclusion, teacher perspectives, school counselling and resilience.

Kakli Gupta is a clinical psychologist practising in Bengaluru. She holds a doctorate in Clinical Psychology from the California School of Professional Psychology, San Francisco, USA. Dr. Gupta works with adults, children, adolescents, couples and families. She is involved in the training as well as personal development of many mental health professionals in Bengaluru. Dr. Gupta has been using video conferencing in her practice for the last 7–8 years as a way to provide therapeutic services to long distance as well as physically challenged clients.

Rathna Isaac has been in private practice for 10 years in Bengaluru. She completed her M.Phil. and Ph.D. from NIMHANS, Bengaluru. Her doctoral work examined the process and outcome of marital therapy with distressed couples. She has held teaching positions in Bengaluru University and Montfort College and held a position as Junior Scientific Officer at NIMHANS. Currently, she is an external consultant for the Couple and Family Therapy programme at Parivarthan Counselling and Training Centre. She has created and conducted basic and advanced training modules on couple therapy and also provides ongoing group supervision for couple counsellors. She has conducted several training workshops for counsellors and psychotherapists; has publications in the areas of couple relationships, ethical issues and psychotherapy; and is deeply interested in the psychotherapy process.

Sanghamitra Iyengar is the founder trustee of Samraksha, a nongovernment organization working on developmental issues in North Karnataka, with a special focus on HIV and AIDS, for the past 23 years. This has included a long-term engagement with women in sex work and their life issues beyond just health. She is a social worker by training, with a Masters in Social Work from Bengaluru University and an M.Phil. in Psychiatric Social Work from NIMHANS. Her practice has spanned disability, mental health, HIV and AIDS and palliative care. Her areas of interest include strength-based approaches, community strengthening, gender, qualitative research, capability building and end-of-life care.

Anuradha Kapoor is the Founder and Director of Swayam, a feminist organization committed to advancing women's rights and ending gender inequality and violence against women, established in May 1995. Anuradha is an activist, trainer and consultant for women and human rights organizations, academic institutions and UN agencies and has contributed to research and manuals addressing violence against women. She is actively involved in advocacy of women's rights in India and abroad. She keenly advocated for the introduction of the Protection of Women from Domestic Violence Act (2005) and its effective implementation. She has also contributed to numerous other legal and policy initiatives, including the Alternative Report for India to the CEDAW (Convention on Elimination of all forms of Discrimination Against Women) Committee in 2014. Anuradha is both an Ashoka Fellow and an Eisenhower Fellow and a member of the Civil Society Advisory Group for the UN Women's Multi Country Office for India, Bhutan, Sri Lanka and the Maldives.

J. Lakshmi is currently pursuing a Ph.D. in Clinical Psychology at NIMHANS. She holds B.Sc. and MA degrees in the field of psychology. She completed her training in M.Phil. in Clinical Psychology from NIMHANS in 2012. Her research interests include cognitive behaviour therapy, metacognitive therapy and mindfulness-based approaches to managing psychiatric disorders.

Suresh Bada Math is Additional Professor, Department of Psychiatry, NIMHANS. He completed his MD in Psychiatry in 2001 from NIMHANS and DNB in Psychiatry in 2005. He holds a Postgraduate Diploma in Medical Ethics and Law (PGDMLE) and Postgraduate Diploma in Human Rights and Law (PGDHRL) from National Law School of India University (NLSIU), Bengaluru. He also holds a Ph.D. in Law from NLSIU. In addition to his clinical and research work, he is in-charge of the Free Legal Aid Clinic and is the Nodal Officer of the Telemedicine Centre at NIMHANS. His research interests include obsessive compulsive disorder, forensic psychiatry, human rights and disaster mental health. He has recently completed a research project examining mental health and substance use problems in prisons.

R. Raguram is Professor and Head, Department of Psychiatry, Kempegowda Institute of Medical Sciences, Bengaluru. He was formerly Professor at the Department of Psychiatry at NIMHANS for over 25 years, where he was instrumental in setting up a structured programme for training in psychotherapy for residents. He was also the coordinator for academic programmes in the department. He is a well-known teacher and researcher with a special interest in the area of cultural psychiatry. He served on the Board of Governors of the World Association of Cultural Psychiatry for two terms and recently guest-edited a special issue of the *Journal of the World Association of Cultural Psychiatry* which focused on mental health issues relevant to the Indian context. He has over 50 publications in national and international journals as well as several awards to his credit.

Vijaya Raman is Additional Professor of Clinical Psychology in the Department of Psychiatry at St. Johns National Academy of Health Sciences, Bengaluru. Dr. Raman obtained her M.Phil. and Ph.D. degrees from NIMHANS. She has been instrumental in establishing the Consultation-Liaison Child Mental Health Services in St. John's and has been associated with it since 1996. She is actively involved in conducting workshops for paediatricians, teachers and parents on increasing awareness of child and adolescent mental health issues. Her areas of research interest include psychological aspects of chronic medical illness in children and adolescents, children and adolescents at high risk, shyness in children and its ramifications, implications of child abuse and child and adolescent outcomes of children with parents who have psychiatric illness. She has a special interest in psychotherapy with children and has extensive clinical experience in the area of play therapy.

Ketki Ranade is Assistant Professor, School of Social Work, at the Tata Institute of Social Sciences (TISS). She recently completed her Ph.D. titled, "Growing Up Gay in a Heterosexually Constructed World", from TISS. She has worked in the mental health sector for over a decade in the capacity of a mental health service-provider, researcher and advocate. Her recently completed research projects look at developing gay affirmative counselling practice and at familial responses to same-sex sexuality in a family member.

Priya Saxena is Assistant Professor in the Department of Clinical Psychology, LGBRIMH, Tezpur. She completed her M.Phil. in Clinical Psychology and Ph.D. from NIMHANS. Her research interests are in the areas of psychotherapy, health psychology and positive mental health.

Shekhar Seshadri is a graduate of Maulana Azad Medical College, Delhi and a postgraduate in psychiatry from NIMHANS. He is currently Professor, Department of Child and Adolescent Psychiatry, NIMHANS. Besides working with child and adolescent mental health including developmental disabilities, he is actively involved in working with areas of gender, sexualities, masculinities, violence/trauma and abuse, children in difficult circumstances, juvenile justice, experiential methodologies, life skills education, school and community mental health initiatives, forum theatre and qualitative research.

Ellen Mrinalini Shinde is Clinical Director, one to one help, an employee assistance programme, and is responsible for supervision and quality control of all counselling activities. She is involved in recruitment and training of counsellors, provision of ongoing consultation and supervision for managing sensitive or critical incidents with employees, and training managers in identifying and responding to employee distress. She has a Masters in Psychology with a specialization in clinical psychology as well as a B.Ed. degree and a diploma in counselling. She has over 24 years of experience in mental healthcare and professional counselling and has previously worked as Senior Counsellor at the Medico-Pastoral Association. She has intensive training in suicide prevention and was associated with running the suicide prevention tele-helpline, Sahai. In addition, she has extensive experience in conducting training programmes on diverse topics across large Indian and multinational companies.

Ananya Sinha is a Ph.D. research scholar and a Senior Research Fellow (University Grants Commission), currently pursuing her Ph.D. from the Department of Clinical Psychology, NIMHANS. Her doctoral work delves into the ethical aspects of psychotherapy and counselling. She has completed her M.Phil. from the Institute of Human Behaviour and Allied Sciences (IHBAS), Delhi. She has served as a Clinical Psychologist in Dumdum Central Correctional Home, West Bengal and has been instrumental in creating and conducting individual and group therapy modules for the inmates in the correctional setting. She has also held teaching positions in the University of Calcutta and the Indian Psychoanalytical Society. Her areas of interests include ethical practice in psychotherapy and counselling, cognitive and emotional processing of traumatic events, social skills training, sexual minorities and borderline personality disorder. She has been involved in conducting workshops and has presented papers related to her areas of interest in both national and international conferences.

Sobhana H. is Associate Professor (I/C) in the Department of Psychiatric Social Work, LGBRIMH, Tezpur, Assam. She has an M.Phil. in Psychiatric Social Work and Ph.D. from NIMHANS. She has undergone a Fogarty Indo-US Post-Doctoral training programme on CNCD at the University of Florida, Gainesville, USA. She is a trained family and couple therapist and is in-charge of the Family Therapy Centre at LGBRIMH. Her research interests are marital and family-related areas, children and adolescent mental health, violence, gender issues, women and mental health, geriatric mental health, disaster, displacement and mental health, and psychosocial interventions.

Shoba Srinath is Senior Professor in the Department of Child and Adolescent Psychiatry, NIMHANS, with more than three decades of academic and clinical experience. She has been the Head of Department of Psychiatry and later the Department of Child and Adolescent Psychiatry as well. She has been the former Dean of the Division of Behavioural Sciences and has vast experience in matters related to biomedical ethics, as she has served on the ethics committees of NIMHANS and other organizations.

B. Vinay is Assistant Professor in Oxford Medical College, Bengaluru. He has an MD in Psychiatry from NIMHANS and an MBBS degree. He recently completed a postgraduation diploma in medical law and ethics from National Law School of India University, Bengaluru. He has previously worked as a senior resident in adult psychiatry at NIMHANS. His key areas of research interest include adult psychiatry, addiction medicine and forensic psychiatry.

Chapter 1
Navigating the Ethical Landscape: Critical Issues in Practice and Training

Poornima Bhola and Ahalya Raguram

A man without ethics is a wild beast loosed upon the world.
—Albert Camus.

Abstract Counsellors and therapists must provide a safe space and walk the ethical line during their psychotherapeutic interactions with vulnerable clients. Ethical practice is informed by several philosophical positions; deontological, utilitarian, the widely used principle-based framework, feminist care ethics and other post-modern perspectives. Expanded frameworks of ethics encompass context, culture, and personal values and embrace the complexity of ethical dilemmas encountered in the therapy room. Research has identified common ethical dilemmas; confidentiality and its limits, boundary violations, therapist self-disclosure, but this has also raised questions about the intersections of culture and ethics. Various professional bodies have delineated ethical codes and guidelines but there are inevitable gaps in their translation to the arena of therapeutic practice. The application of ethical principles may be influenced by a range of client, setting, therapy (theoretical orientation or therapy modality), and organisational variables as well by extant legal frameworks and socio-political contexts. Training in ethics for mental health professionals is inconsistent and often inadequate. A critical evaluation of ethical codes, innovative training methodologies, the need for training in ethical decision-making models, and the value of therapist self-reflection are discussed, with a special focus on the Indian context.

Keywords Ethical decision-making · Ethics training · Psychotherapy · Counselling · Ethical dilemmas · Philosophy of ethics

P. Bhola (✉) · A. Raguram
Department of Clinical Psychology, National Institute of Mental Health and Neuro Sciences, Bengaluru 560029, India
e-mail: poornimabhola@gmail.com

A. Raguram
e-mail: ahalyaraguram@yahoo.co.in

P. Bhola and A. Raguram (eds.), *Ethical Issues in Counselling and Psychotherapy Practice*, DOI 10.1007/978-981-10-1808-4_1

Navigating through the landscape of psychotherapeutic interactions with vulnerable clients requires that therapists be guided by an ethical compass. Client safety and wellbeing are paramount in this human encounter with vulnerable persons, which often occurs behind the closed doors of a 'therapy room'. The onus is on the counsellor or therapist to provide this safe space and walk the ethical line.

Ethical frameworks are interpreted and implemented within professional and cultural contexts. In India (and this may be relevant to other contexts as well), clients are typically less aware of their rights and come from diverse socioeconomic and subcultural contexts (Isaac 2009). Training in ethics for mental health professionals is inconsistent, often inadequate and sometimes absent, and the levels of accountability in the helping professions are low. With the pressures to increase human resources and reduce mental health inequalities in the country, there has been a closer examination of professional training competencies and practice issues in the field.

Ethical competency—the knowledge and understanding of ethics codes, the ability to recognise, analyse and resolve ethical situations and to understand oneself as being responsible for one's own actions (Sporrong et al. 2007), is the cornerstone of professional development in the helping profession. Ethical practice is informed by diverse philosophical positions (Beauchamp 2001), each of which has differing implications for the ethical practice of psychotherapy. As practitioners, we may be concerned with the practical aspects of ethics in real-life therapy scenarios but an understanding of the various complementary and competing philosophical theories of ethics can help us think more deeply and critically about our experiences.

1.1 Philosophical Positions in the Field of Ethics

As a branch of philosophy, ethics seeks to define the concepts of right and wrong, good and evil, virtue and vice and guides our choice of actions in complex situations.

The *deontological* approach by Immanuel Kant speaks of universal imperatives like moral duties and obligations with actions being intrinsically good or bad while the *utilitarian* perspective by John Stuart Mill takes a cost-benefit approach to judge actions as right or wrong depending on their consequences.

A combination of these two perspectives is reflected in the widely used four pillars of *principle-based ethics* (Beauchamp and Childress 2001): beneficence (acting with the person's best interests in mind), non-maleficence ("above all, do no harm"), justice (an emphasis on fairness and equality among individuals) and respecting autonomy (regard for the person's right to make a reasoned informed choice).

In most situations where principles conflict, there is no clear algorithm about which one should take precedence. Many ethicists see non-maleficence as the primary ethical obligation when two ethical principles are in conflict (Kitchener 1984). The principle-based approach is often critiqued for the limited room to consider the circumstances in which the choice must be made (Kitchener 1984). This approach also does not articulate the role of intuition or the impact of personal values in the decision-making process.

Moving away from rules or consequences, Aristotle's *virtue ethics* focuses on an individual's personal qualities, reflections and behaviours as key elements in ethical thinking. Feminist and other post-modern approaches in ethics provide another counterpoint to the principle-based stance of looking at 'oughts' and 'shoulds' within a universalised ethical paradigm. These approaches embrace diversity, uncertainty and varied local, cultural and historical factors that interact in making ethical choices (McNamee 2009). The feminist *care ethics* position uses a more inductive process of reaching ethical conclusions and rethinks the deontological and utilitarian philosophies which are viewed as more deductive or mathematical (Gilligan 1982). It sees people as interdependent and includes perspectives like responsibility, empathy and compassion in making ethical choices. While the principle-based ethics of justice and the ethics of care are often viewed as representing two opposite and irreconcilable poles (Botes 2000), a careful reading reveals that Gilligan did not see care ethics as replacing justice-based ethics but advocated a flexible shifting between these two approaches (Proctor 2014).

While the debates about these different moral epistemologies continue, this has given rise to a more expanded framework of ethics. There is space for debate between *ethical absolutism*, which adopts a black or white lens and the *ethical relativism* view, which sees ethical judgements as true or false relative to a particular context. Variations in individual moral philosophies have been described in terms of four ethical outlooks (Forsyth 1980). *Situationists* are individuals who advocate a contextual analysis of morally questionable actions; *absolutists* use universal moral principles to formulate moral judgments and feel that some ethical absolutes are important enough to be included in any code of ethics; *subjectivists* argue that moral judgments should depend primarily on one's own personal values; and *exceptionists* agree that exceptions must sometimes be made to moral absolutes.

The philosophical position adopted by an individual or within a training programme influences ethical decisions in real-life therapy scenarios. Therapeutic encounters engender the need for making multiple ethical decisions but these may not always be experienced as dilemmas by the therapist. For instance, a therapist may be careful to avoid discussing confidential client revelations with friends, family members or even colleagues and this may be well integrated into the practitioner's professional identity, role and functioning (Hill et al. 1998). Nevertheless, the experience of challenging ethical paradoxes or dilemmas, 'a problem for which no course of action seems satisfactory' (Kitchener 1984), is an inevitable and recurrent occurrence for all counsellors and therapists.

1.2 "*To Be or Not to Be*": Ethical Dilemmas in the Therapy Room

The experience of an ethical conflict arises "because there are good, but contradictory ethical reasons to take conflicting and incompatible courses of action" (Kitchener 1984). The conflict can arise from a clash between any combination of the professionals' personal values or feelings, ethical codes, law, aspects of personal or professional loyalties and clinical knowledge and experience (Hill et al. 1998). These dilemmas involve uncertainties about what is right and wrong and go beyond 'therapeutic dilemmas' about which interventions are most appropriate or effective in a given situation.

Surveys on ethical beliefs and practices can help identify the common dilemmas experienced by therapists and isolate the areas where there are differing interpretations of ethical maxims. It is particularly important to understand the controversial zones where the professional community is split in their views. Psychotherapists may experience the pressure to be wise, all-knowing and even infallible and personal experiences of troubling ethical dilemmas or transgressions could remain concealed. Safe professional spaces to disclose and discuss ethical issues are often limited and practitioners may well be ambivalent or uncertain in certain areas due to lack of familiarity and knowledge.

Key findings from these surveys have direct implications for training and for the expansions or revisions of professional ethics codes (Gibson and Pope 1993). Many of the initial surveys were conducted in North America and used either vignettes, open-ended questions or a list of specific behaviours to be rated in terms of how often practitioners engaged in this behaviour and their evaluation of how ethical the behaviour was (Haas et al. 1986; Gibson and Pope 1993; Pope and Vetter 1992).

One the earliest surveys (Haas et al. 1986) of psychologists in the United States of America identified the most prominent ethical dilemmas as related to confidentiality, concerns about own or colleague's professional competence and the appropriate response to a colleague's unethical behaviour. Confidentiality is a core professional value and is seen as essential to develop and maintain a trusting therapeutic relationship but decisions about disclosure appeared to pose the most confusion. Psychologists were clear about the limits of confidentiality and how to handle their duty to warn regarding a client's potentially violent behaviour but experienced uncertainties about maintaining confidentiality within the family. Findings from another survey (Pope and Vetter 1992) also pointed to confidentiality as the most frequently encountered ethical dilemma. Psychologists described challenging incidents related to actual or potential risks to third parties, child abuse reporting and disclosure of HIV status. They grappled with decisions about sharing sensitive client information with supervisors or administrators or in the context of group or couple therapy,

Decisions about disclosure and non-disclosure do not always align with ethics codes or legal requirements and clinicians may commit *intentional* violations on occasion, influenced by their own strongly held values or to protect client welfare (Pope and Bajt 1988). Many violations of confidentiality are often *unintentional* (Pope et al. 1987) and together these represent an area that is perhaps most challenging for practitioners. The complex and sometimes competing demands of professional ethics guidelines, client protection, laws and risk management means that psychologists might have 'difficulty seeing the ethical forest for the legal trees' (Fisher 2008, p. 6). Confidentiality is often conditional rather than absolute, with some disclosures being legally mandated or seen as ethically appropriate. Given this, it is critical that any possible exceptions to confidentiality are discussed with clients at the very outset of the interaction. Practitioners may view this as "clinically inconvenient" (Fisher 2008, p. 4) and as something that could come in the way of building a therapeutic alliance, but the client's right to provide informed consent is unassailable.

Therapists have a circumscribed role in their clients' lives and need to work within the frame of this fiduciary relationship. The avoidance of potentially harmful or exploitative dual relationships with clients (e.g. having a sexual or business relationship with a client) is outlined in most professional ethics codes. Nevertheless, surveys have indicated that ethical dilemmas often stem from situations where the delineation of clear therapeutic boundaries is less clear (Borys and Pope 1989; Pope and Vetter 1992). Therapists reported uncertainties in a range of circumstances; for instance, being both a therapist and supervisor for a person, employing a former client, accepting a gift from a client, accepting a client's invitation for special occasion or providing therapy to a client's friend, relative or lover (Borys and Pope 1989).

There was little ambiguity about sexual involvement with a client; with almost all therapists (98.3 %) viewing this as never ethical (Borys and Pope 1989). Interestingly, the patterns were different concerning sexual activity with a client post-termination; 68.4 % of therapists considered this as absolutely unethical and the remaining viewed this as ethical to different degrees under varied circumstances (Borys and Pope 1989). Despite strong prohibitions against sexual boundary violations, these incidents continue to occur within the profession. Although research is limited in the Indian setting, a survey revealed a worryingly high proportion of psychiatrists and psychologists who reported awareness of sexual boundary violations by a mental health professional (Kurpad et al. 2010). More often than not, the allegations were not investigated and the majority felt the need for specific guidelines concerning both sexual and non-sexual boundary violations in India.

Practitioners often experience uncertainties about where to draw the line with respect to the extent of therapists self-disclosure or when clients bring gifts. Most ethics codes do not provide explicit guidance on these behaviours and such boundary crossings may be viewed as relatively benign. Gutheil and Gabbard (1993) cautioned against the 'slippery slope' where small steps across the line could eventually lead to more serious boundary violations which cause harm to the client.

Research findings point to the need for clearer distinctions between potentially harmful dual relationships and accidental, incidental contacts, for instance, encountering a client at a social occasion. Ethical codes also need to address the realities of professionals working in rural and small communities for whom dual relationships may be inevitable (Pope and Vetter 1992) and where confidentiality is more difficult to ensure (Helbok 2003). The profession can address the challenge of drawing boundaries either by constructing rigid fences or more flexible frames that protect both the therapist and the client.

Research points to less frequently encountered ethical dilemmas; for instance, practising in areas outside one's competence, recordkeeping and documentation; but these are by no means less challenging for practitioners when they do occur.

Surveys provide a snapshot of the most frequently encountered ethical dilemmas but may not capture the complexity of real-life clinical scenarios where the answer is often not 'yes' or 'no' but 'it depends'. Future research can incorporate a qualitative perspective and increase our understanding of ethical challenges in the field.

The context of therapy work can give rise to unique ethical issues, influence what we experience as a dilemma and even how we choose to respond. For practitioners trying to navigate the ethical landscape, the cultural context, the therapy setting (e.g. private practice, workplaces, schools, rural communities, online), client characteristics (e.g. children, survivors of sexual violence, persons with lesbian, gay or bisexual orientation), therapy modality or orientation (e.g. couple therapy) and personal or professional characteristics of the therapist (e.g. gender, years of experience) can all influence the ethics of interactions with clients.

1.3 Ethics in Context

Discussions about the intersections of ethics and culture are moving to the forefront in current times. Surveys of ethical dilemmas across countries and cultures have indicated both common and divergent findings. Research in Australia (Sullivan 2002) indicated a trend of strong agreement on common and ethical behaviours and in terms of difficulties related to fees and financial transactions with clients, similar to earlier surveys in North America (Gibson and Pope 1993; Pope et al. 1987). Both South African clinical psychologists (Slack and Wassenaar 1999) and clinical psychology trainees in India (Bhola et al. 2015) experienced confidentiality and the negotiation of boundaries as the most pressing ethical dilemmas. Miller et al. (2012) proposed that there are many cultural universals and that cultural factors may be most influential on those kinds of behaviours that are perceived as relatively less serious.

Social structures are different across countries and this can reflect in professional ethical beliefs; for instance, Chinese practitioners expressed different beliefs about the need to obtain informed consent (Zhang et al. 2007). In interdependent cultures like India, the involvement of family members in the treatment process or the client referring to the therapist (often seen as a benevolent authority figure or elder) as a family member is more common. This can influence the construction of client autonomy within an egalitarian therapy relationship and the interpretation of guidelines related to confidentiality and boundaries.

Dual relationships may be viewed differently in traditional collectivistic societies. Two surveys in China revealed that between one-third to half of the counsellors and psychotherapists provided therapy for friends, students, supervisees or employees and did not necessarily see this as an ethical transgression (Zhao et al. 2009; 2011). Viewing this through the dominant Western lens would suggest that these deviations warrant increased professional inputs and supervision. A more culturally sensitive approach might educate trainees on how best to maintain boundaries in the context of a pre-existing relationship between therapist and clients, rather than trying to construct rigid boundaries (Zhao et al. 2011).

While many ethics codes mention the importance of therapist multicultural competence, a larger question concerns the cultural universality versus specificity debate about ethics codes themselves. Some ethical guidelines like the American Counseling Association (ACA 2005) discuss the importance of the cultural context with respect to ethical guidelines. Therapists are required to interpret certain behaviours, e.g. gift giving, in relation to the clients' cultural values and background. While a particular behaviour may be ethically questionable in one culture, it could be completely acceptable in another. Some of the ethical codes in India do not discuss the issue of accepting gifts from clients, even though practitioners often experience difficulties in dealing with such behaviours (Bhola et al. 2015; Kurpad et al. 2010). The Indian Psychiatric Society (Ahuja 2006) prohibits accepting gifts; while this blanket prohibition may serve as a protection against potential exploitation, it may also be at odds with cultural realities. There is a need for open discussions and nuanced debates about the cultural universality or validity of the ethical guidelines derived from international frameworks.

Ethical guidelines may need to be modified to reflect cultural realities which are different from those in individualistic cultures. Zhao et al. (2011) suggested that the Chinese *Code of Ethics for Clinical and Counseling Practice* would benefit from modifications so that is more true to the traditional Chinese culture with its emphasis on interpersonal relationship networks, mutuality and reciprocity, as well as hierarchy and respect for authority.

While the approach of ethical relativism respects diversity, it cannot entirely circumvent or transform the core ethical guidelines formulated for the profession. The rapidly changing socio-cultural processes and the socio-political inequalities, particularly in the developing nations like India, also make it difficult to have a static understanding of ethical codes.

Many of the chapters in this book illustrate the unique ethical challenges that arise from working in diverse settings and with varied client groups. Therapist-client interactions in the digital space, through emails, instant messaging or through video-enabled online sessions, have thrown up a set of contemporary ethical concerns.

Research in other countries has indicated much consensus in the ethical beliefs held by members of different mental health professions (e.g. psychiatrists, psychologists and social workers) but also revealed differences related to select behaviours (Borys and Pope 1989; Pomerantz et al. 1998). The lack of research in India makes it difficult to comment on any variations in the ethics of practice across the helping professions.

A striking gap is the lack of information about client perspectives and experiences. One study reported that patients and lay persons tend to be less strict than therapists regarding issues of boundaries (Fennig et al. 2005). The onus is then even more on the therapist to ensure that the boundaries and frames of the therapeutic encounter are adhered to, particularly in countries like India where consumers have limited awareness about client rights and professional ethics.

Clearly, there are potential ambiguities in ethical viewpoints and practitioners need to make continuous judgments about how to respond to the inevitable ethical dilemmas. Professional ethical frameworks can provide some guideposts as we navigate the ethical landscape but do they always provide a clear road map?

1.4 Ethical Guidelines: A Clear Road Map?

Most ethical guidelines follow similar basic values and principles but tend to vary in the degree of detail provided. The commonly cited guidelines from the American Psychological Association (American Psychological Association 2010), the British Association of Counselling and Psychotherapy (British Association of Counselling and Psychotherapy 2010), The American Psychiatric Association (2013) and the National Association of Social Workers (NASW) (2008) outline ethical principles and expected standards of conduct in an elaborate manner.

In the Indian scenario, the Indian Association of Clinical Psychologists (IACP) and the National Academy of Psychology (NAOP) provide ethical guidelines for the teaching, training, practice and research in the field of Psychology. The Indian Psychiatric Society prepared a code of ethics for psychiatrists in 1989 (Ahuja 2006). The psychiatric social work profession in India refers to the National Association of Social Workers (NASW) Code of Ethics developed in the United States of America.

The Indian Association of Clinical Psychologists has initiated work on revisions to their Ethics code developed in 1995. This is an opportune time to review and revisit some critical questions about ethical guidelines and implications for therapy training and practice in India. The recent draft version of the Indian Association of Clinical Psychologists Code of Ethics (IACP 2015) has added many details and seems to mirror many sections of the APA code which is strongly rooted in the principle ethics framework (Knapp and VandeCreek 2004). In comparison to the American Psychological Association (APA) code, the IACP code is less comprehensive and does not include specific details concerning assent, multiple relationships, sexual relationships with clients, fees, forensic issues, documentation and public statements and advertising.

The National Academy of Psychology (NAOP 2010) guidelines elaborate on the values of caring, respect, integrity and professional and scientific responsibilities to society. However, specific ethical guidelines for psychotherapeutic practice are not outlined in the document.

Appendix A provides a brief description of select international and Indian ethics codes along with information about how to access them.

Practitioners may have individual interpretations of the codes as well as of the term ‘ethical’ (Walker 1994) and may rely on their own value systems in situations where ethical guidelines are unclear (Bersoff and Koeppl 1993). The traditional view of the therapist as a ‘blank slate’ in the scientific enterprise of therapy, with all personal values kept firmly outside the therapy room, persists despite the literature suggesting that therapists and therapy cannot be completely value-free (Bergin et al.

1996). Ethical codes tend to caution therapists against imposing their values on clients but these are brief directives which do not fully acknowledge the way values and ethics are intrinsically linked.

The application of ethical principles may be influenced by a number of factors such as client variables, theoretical orientation, setting or organizational variables. For example, specific ethical dilemmas may arise in working with minor children and adolescents and must be made explicit in the ethical guidelines in India (de Sousa 2010). Similarly, unique and complex ethical issues may arise in couple and family therapy and require clarification in professional ethical codes (Margolin 1982).

Critical evaluation of professional guidelines is essential to assess and address inevitable gaps between ethical principles and their translation in the arena of therapeutic practice. Professional ethical guidelines are not written in stone and may need changes and revisions as society metamorphoses and evolves. How often are these amendments required and what should the decision-making and consensus process involve?

There are some lessons from examining the process of the ten amendments of the American Psychological Association Ethics code from the first version in 1953. The code evolved from reviewing more than 1,000 incidents involving ethical decision-making by members of the APA (Fisher 2012). The 2002 revision included open calls for comments including diverse segments like legal experts, ethicists, human rights activists and consumers. While it would be advisable to review ethical codes with specified time frames, there should be room for unscheduled amendments as well. In 2010, the APA reviewed the controversial clause which allowed dispensing with professional ethics, if these conflicted with 'law, regulations, other governing legal authority'. This change was aimed at protecting against potential conflicts concerning human rights violations, for example the involvement of psychologists in acts of torture in war or military contexts.

Contemporary challenges related to therapist–client interactions and interventions through the digital media and the Internet may not be adequately addressed by all professional ethical codes. These include the possibility of misunderstanding and miscommunication without face to face contact; electronic confidentiality concerns and boundary issues, difficulties in response to crisis situations and possible technological problems and interruptions. The availability of both professional and personal information about therapists on the World Wide Web and on social networking sites is a new arena of potential challenges. Although the basic ethical frameworks may be similar, additional issues in e-therapy have been addressed in a separate section in some codes (American Counseling Association 2005) or in a more detailed set of specific guidelines (International Society of Mental Health Organisation 2000).

In contrast to most international codes, the guidelines of the Indian Association of Clinical Psychologists and the Indian Psychiatric Society have not been revised since the time of their inception more than two decades ago. The continued relevance of ethics guidelines will depend on responsiveness to change with fixed timelines for review and the scope for out-of-turn amendments when required.

1.5 Intersections Between Ethics and the Law

Ethics invariably intersect with legal frameworks and the changing socio-political climate. On one hand, changes in laws might call for corresponding modifications in ethical codes. For example, The Protection of Children from Sexual Offences Act, 2012 was notified in June 2012 and calls for mandatory reporting of child sexual abuse in India, with legal sanctions for non-compliance. This has significant implications for counsellors and therapists practising in a variety of settings in the country and calls for reflection on possible inclusion in professional ethics codes.

On the other hand, it is possible that the professional community might have views that differ from extant laws in the country. For example, the American Psychiatric Association made position statements that had significant implications for ethical decision-making in psychotherapy practice; one against reparative therapies for homosexuality (Anton 2010) and another resolution stating that denying same-sex marriages was unfair and discriminatory (American Psychological Association 2011). The viewpoint on same-sex unions was at variance with laws in many states in the United States of America. In India, the Indian Psychiatric Society and the Karnataka Association of Clinical Psychologists have made official statements against the concept of homosexuality as a mental illness (Orinam 2014) and against reparative treatments for homosexuality (Hemchand 2016) but these have not been integrated into the ethics codes. Other professional organizations in the country have been silent on this issue and this must be viewed in the backdrop of Section 377 of the Indian Penal Code which criminalizes same-sex intercourse.

Trainees and practitioners in the country need to be aware of existing laws that may have implications for the ethical and legal conduct of therapy. Some of these include the Domestic Violence Act, 2005; the Narcotics and Psychotropic Substances Act, 1987; Protection of Children from Sexual Offences Act, 2012 and The Juvenile Justice (Care and Protection of Children) Act of 2000. According to Section 309 of the Indian Penal Code, an attempt at suicide is punishable by a fine and/or imprisonment until 1 year. This may inhibit persons from seeking help for associated mental health concerns, if any. Medico-legal processes need to be carried out by a practitioner working in a hospital setting and this throws up challenges in straddling the ethical and legal obligations towards a vulnerable client. Although there have been efforts to remove this section or introduce some limitations through the proposed Mental Health Care Bill (Narayan and Shekhar 2015), the process is ongoing. The provisions of the law may be amended from time to time and mental health practitioners may even be involved in discussions leading up to revisions of laws or formulations of new ones.

Ethical issues related to therapeutic competence are complicated by uncertainties regarding registration and licensing of counsellors and therapists in India. Isaac (2009) highlighted concerns about reporting, accountability and legal sanctions for ethical transgressions by psychotherapists in India. However, recent developments like the Right to Information Act, 2005 and the Consumer Protection Act, 1986 have provided avenues for information-seeking and redressal. This has increased concerns about documentation, confidentiality and related ethical issues among practitioners and is accompanied by the possibility of a shift to more defensive practice patterns. In a recent case involving a request filed under the Right to Information Act, for access to an estranged spouse's confidential medical records, the Chief Information Commissioner ruled in favour of access in the larger 'public interest'. Nair (2015) discussed the possible violations of professional ethics as well as an individual's constitutional rights in such scenarios and suggested some restrictions on the RTI act through legislative changes.

Avasthi and Grover (2009) examined complex dilemmas related to documentation of therapeutic encounters in terms of medico-legal ramifications in India. Effective documentation of therapy sessions can serve as a guide to practice and may even protect the clinician in the event of any litigation. On the other hand, the sensitive personal information may be open to disclosures in a court of law and this has its own ethical ramifications. Avasthi and Grover (2009) highlighted the lack of clarity about privileged communication and the limits of disclosure in the Indian law.

Is there any enforcement value to ethical codes? How much is it meant to monitor professional behaviour? Recent research in India revealed that although practitioners were aware of therapist–client sexual boundary violations committed by colleagues (Kurpad et al. 2010), there was little clarity on reporting mechanisms within the professional organization and in the larger legal system. The processes for reporting colleagues' ethical misconduct are ill-defined in India and even where provisions exist on paper, they may be rarely used.

Work in forensic settings may throw up ethical challenges that differ from those in traditional practice. Psychologists are increasingly being called upon to conduct assessments and offer expert opinions on psycho-legal issues. Working with children in conflict with the law requires adequate knowledge about the legal framework for juvenile justice in India (Jacob et al. 2014). Contemporary professional standards and guidelines should anticipate the interlinked legal, professional, moral and ethical factors which can inform practice in the forensic speciality area.

Ethics and the law are closely linked but not always synchronous and there is a need for more open communication between the disciplines of mental health and the law. Mental health trainees and professionals need to be aware of legal frameworks that can influence their practice in the states or countries where they work.

1.6 Ethics Training: How and When to Teach (and Learn) Ethical Practice?

> Ethics is the study of right and wrong but is often taught as the study of wrong. (Handelsman et al. 2005).

Training in ethics during the early stage of the professional journey is often inadequate in both content and format. In most Indian training contexts, ethics is either neglected or taught as a stand-alone class, with ethical codes taught as if they are the essential navigation guide through a landscape that is in reality far more complex and uncertain. Ethical codes may be 'necessary but not sufficient' (Barnett et al. 2007) and are a small component of ethics training. In a recent Indian study (Bhola et al. 2015), clinical psychology trainee therapists reported mixed perceptions about the utility of professional codes in resolving ethical dilemmas. Some of their concerns included the lack of specificity or clarity in the guidelines, poor responsiveness to unique contextual aspects of therapeutic situations, and the conflict between guidelines and other values. The aim of ethics training needs to go beyond a set of rules telling you '*what to do (or not)*' and instead inculcate the process of '*how to think*', using ethical decision-making models.

Ethical decision-making models provide a framework and a stepwise process for navigating complex and sometimes contradictory variables that arise during therapy. There are a range of ethical decision-making models which may be philosophically based, practice-based or relevant to certain specialities (Cottone and Claus 2000). Most models offer a concrete flow chart of steps of rational decision-making; beginning with identifying the problem, the stakeholders, applicable codes, principles and laws; and using a cost-benefit analysis to choose between different emerging options. Feminist ethical decision-making models argue against this objective and intellectualized approach and seek to take into account the 'person of the therapist' (emotional-intuitive responses, personal experiences, characteristics and values), situational, contextual and cultural elements as well as the client's perspective and reactions. The feminist model by Hill et al. (1998) includes a series of self-reflective questions at every step, reflects both emotional and cognitive aspects and can be a useful resource for practitioners. Corey et al. (2011) have integrated perspectives and questions influenced by feminist models, social constructionist approaches and transcultural integrative model of ethical decision-making to propose a stepwise process for thinking through ethical dilemmas: Identify the problem or dilemma, Identify the potential issues involved, Review the relevant ethics codes, Know the applicable laws and regulations, Obtain consultation, Consider possible and probable courses of action, Enumerate the consequences of various decisions; Choose what appears to be the best course of action. An expanded 17-step ethical decision-making model proposed by Pope and Vasquez (2016) also includes useful suggestions for documentation and review of the intended or unintended consequences of the decision and possible implications for future preparation, planning and prevention.

Ethical questions may not have a single, invariant answer, and some have even asserted that, "there are no answers, only choices" (Gray and Gibbons 2007). Ethical decision-making approaches and therapist self-awareness can be key tools that help practitioners understand, evaluate and select among the choices (Evans et al. 2012).

The broader objectives of ethics training are really to learn how to think and work with debatable positions, shift positions and tolerate the grey area of uncertainty as the practitioner work out what seems best. An interactive learning environment with creative exercises emphasizing therapist reflection is what can enhance professional growth and training in applied ethics. Looking beyond the therapy room, debates on controversial studies such as the Stanford prison experiment and Milgram's studies on obedience (Tolich 2014) or issues such as the role of psychiatrists during the Nazi regime (Strous 2007) can yield important ethical lessons. Open discussions on topics like euthanasia, capital punishment, foetal sex-determination, affirmative action for the reservation of seats for women in the parliament, can pave the way for trainees to hold different perspectives in mind and understand the complexities of ethical decision-making. Too often, the teaching of ethics in psychotherapy practice is disconnected from the rest of the training curriculum. The theories of Kohlberg and Gilligan (Blum 1988) can inform our thinking about ethics, but are typically taught as developmental theories of moral reasoning in the undergraduate psychology classes, with little connection to real-life ethical decision-making as a therapist. The different epistemological assumptions inherent in the Kohlberg and Gilligan theories can be effectively used to stimulate classroom discussions. The well-known Heinz's dilemma can be used to think about the way we view client's life stories and also about ethical choices and decisions. In this integrated approach to ethics training, the trainee is given the opportunity to develop the 'inner ethicist' from the early stages of learning and professional development. The scope of ethics training needs to include both practice and research issues in counselling and psychotherapy.

Effective training in ethics can use a range of innovative methodologies and can be informed by practitioner experiences and research findings. These could include discussions using case dilemmas, relevant to the culture and context, collected from trainee and practitioner encounters in their therapeutic work, rather than those distilled from textbooks. Trainees could be encouraged to present multiple viewpoints to the same scenario, with care taken that the facilitator is not seen as the 'expert' who will swiftly provide the 'right answer'. Some other classroom exercises could involve critical analyses of professional ethics codes and their changes over the years, or writing an 'ethics autobiography' where therapists outline how they came to their present notion of what it means to be an ethical practitioner (Handelsman et al. 2005). As we progress along our professional journey and encounter ethical challenges, it is possible that varied experiences change our perceptions and stance. In support of this, Clemente et al. (2011) reported that psychologists who had actually faced a particular dilemma were less strict in judging the inappropriateness of a possible ethical transgression than those psychologists who had not experienced it. Ethics training could include experiential accounts from therapists about these experiences and changes.

'Ethics Rounds', with detailed case discussions of ethical conundrums during therapy sessions with clients, can provide an open and conducive environment for learning in training contexts. Research with clinical psychology trainees in India indicated that supervision or consultation with professional colleagues or peers was the most common strategy for resolution of ethical dilemmas (Bhola et al. 2015). Proactive supervision in a climate of discourse would be most useful for processing ethical concerns. Bernard and Goodyear (2004) caution that, "Waiting for ethical issues to emerge in supervision seems to set up the conditions for crisis training, not ethics training" (p. 71).

Practitioners, particularly those in private practice, may find it difficult to access supervision and consultation for ethical problems in clinical practice. The development of a network of clinical ethics committees (CECs) across the UK has emerged as a practical response to this need and offers specific advice on individual cases and well as ethics education (Edwards and Street 2007).

Many ethical questions might arise at different stages of the professional journey, or as the therapist shifts to newer modes or spaces of work (e.g. online therapy). Continuing education and training in ethics, including workshops, peer supervision networks, the development of manuals and online training modules are directions for the future. The formation of active ethics committees or sections within professional mental health associations in India would lend support to such initiatives.

1.7 Conclusion

Toulmin's (1996) influential thoughts on applied ethics for real-life ethical quandaries, advocate an approach that is "particular not universal, local not general, timely not eternal, and—above all—concrete not abstract" (p. 7). These principles can provide directions for training methodologies and development of codes of ethics that connect with the diverse exigencies of psychotherapy practice in varied contexts. This volume brings together the collective wisdom of practitioners from diverse settings who draw upon their experiences to discuss the complexities of ethical issues that arise in counselling and psychotherapy.

References

Ahuja, N. (2006). *A short textbook of psychiatry* (6th ed.). New Delhi: Jaypee Brothers Publishers.

American Counseling Association. (2005). *ACA code of ethics*. Retrieved December 13, 2013 from http://www.counseling.org/Resources/aca-code-of-ethics.pdf

American Psychiatric Association. (2013). *The principles of medical ethics*. Retrieved May 15, 2016 from https://www.psychiatry.org/psychiatrists/practice/ethics

American Psychological Association. (2010). *Ethical principles of psychologists and code of conduct*. Retrieved December 15, 2013 from http://www.apa.org/ethics/code/principles.pdf.

American Psychological Association. (2011). *Resolution on marriage equality for same-sex couples*. Washington, DC: Author. Retrieved December 15, 2013 from http://www.apa.org/about/policy/same-sex.pdf

Anton, B. S. (2010). Proceedings of the American Psychological Association for the legislative year 2009: Minutes of the annual meeting of the Council of Representatives and minutes of the meetings of the Board of Directors. *American Psychologist, 65*, 385–475.

Avasthi, A., & Grover, S. (2009). *Ethical and legal issues in psychotherapy*. Retrieved December 15, 2013 from www.indianjpsychiatry.org/cpg/cpg2009/article11.pdf

Barnett, J. E., Behnke, S. H., Rosenthal, S. L., & Koocher, G. P. (2007). In case of ethical dilemma, break glass: Commentary on ethical decision making in practice. *Professional Psychology: Research and Practice, 38*, 7–12.

Beauchamp, T. L. (2001). *Philosophical ethics: An introduction to moral philosophy*. McGraw-Hill.
Beauchamp, T. L. & Childress, J. F. (2001). *Principles of biomedical ethics*. 5th Edn. Oxford University Press.
Bergin, A. E., Reed Payne, I., & Scott Richards, P. (1996). Values in psychotherapy. In E. P. Shafranske (Ed.), *Religion and the clinical practice of psychology* (pp. 297–325). Washington, DC: American Psychological Association.
Bernard, J. M., & Goodyear, R. K. (2004). *Fundamentals of clinical supervision*. Boston, MA: Pearson.
Bersoff, D. N., & Koeppl, P. M. (1993). The relation between ethical codes and moral principles. *Ethics and Behavior, 3*(3 & 4), 345–357.
Bhola, P., Sinha, A., Sonkar, S., & Raguram, A. (2015). Ethical dilemmas experienced by clinical psychology trainee therapists. *Indian Journal of Medical Ethics, 12*(4), 206–212.
Blum, L. A. (1988). Gilligan and kohlberg: Implications for moral theory. *Ethics, 98*, 472–491.
Borys, D. S., & Pope, K. S. (1989). Dual relationships between therapist and client: A national study of psychologists, psychiatrists, and social workers. *Professional Psychology: Research and Practice, 20*(5), 283–293.
Botes, A. (2000). A comparison between the ethics of justice and the ethics of care. *Journal of Advanced Nursing, 32*, 1071–1075.
British Association of Counselling and Psychotherapy. (2010). *Ethical framework for good practice in counselling & psychotherapy*. Retrieved December 15, 2013 from http://www.bacp.co.uk/admin/structure/files/pdf/9479_ethical%20framework%20word%20feb2010%20%28revised%29.pdf
Clemente, M., Espinosa, P., & Urra, J. (2011). Ethical issues in psychologists' professional practice: agreement over problematic professional behaviors among Spanish psychologists. *Ethics and Behavior, 21*(1), 13–34.
Corey, G., Corey, M., & Callanan, P. (2011). *Issues and ethics in the helping profession* (8th ed.). Belmont, CA: Brooks/Cole.
Cottone, R. R., & Claus, R. E. (2000). Ethical decision-making models: a review of the literature. *Journal of Counseling & Development, 78*(3), 275–283.
De Sousa, A. (2010). Ethical issues in child and adolescent psychotherapy: A clinical review. *Indian Journal of Medical Ethics, 7*(3), 157–161.
Edwards, S. D., & Street, E. (2007). Clinical ethics committees: a practical response to ethical problems in clinical practice. *Clinical Child Psychology and Psychiatry, 12*(2), 253–260.
Evans, A. M., Levitt, D. H., & Henning, S. (2012). The application of ethical decision-making and self-awareness in the counsellor education classroom. *Journal of Counselor Preparation and Supervision, 4*(2), 41–52.
Fennig, S., Secker, A., Treves, I., Ben Yakar, M., Farina, J., Roe, D., et al. (2005). Ethical dilemmas in psychotherapy: comparison between patients, therapists and laypersons. *The Israel Journal of Psychiatry and Related Sciences, 42*(4), 251–257.
Fisher, M. A. (2008). Protecting confidentiality rights: The need for an ethical practice model. *American Psychologist, 63*, 1–13.
Fisher, C. B. (2012). A code of ethics for psychology. How did we get here? (pp.3–18). In C. B. Fisher (Ed.), *Decoding the ethics code: A practical guide for psychologists* (2nd Edn). California: Sage Publications.
Forsyth, D. R. (1980). A taxonomy of ethical ideologies. *Journal of Personality and Social Psychology, 39*, 175–184.
Gibson, W. T., & Pope, K. S. (1993). The ethics of counseling: A national survey of certified counsellors. *Journal of Counseling and Development, 71*, 330–336.
Gilligan, C. (1982). *In a different voice: psychological theory and women's development*. Cambridge, MA: Harvard University Press.
Gray, G., & Gibbons, J. (2007). There are no answers, only choices: Teaching ethical decision making in social work. *Australian Social Work, 60*, 222–238.

Gutheil, T. G., & Gabbard, G. O. (1993). The concept of boundaries in clinical practice: theoretical and risk-management dimensions. *American Journal of Psychiatry, 150*, 188–196.

Handelsman, M. M., Gottlieb, M. C., & Knapp, S. (2005). Training ethical psychologists: An acculturation model. *Professional Psychology: Research and Practice, 36*(1), 59–65.

Haas, L. J., Malouf, J. L., & Mayerson, N. H. (1986). Ethical dilemmas in psychological practice: Results of a national survey. *Professional Psychology: Research and Practice, 17*(4), 316–321.

Helbok, C. M. (2003). The practice of psychology in rural communities: Potential ethical dilemmas. *Ethics and Behavior, 13*(4), 367–384.

Hemchand, L. (2016). A practitioner's guide to understanding sexual orientation. In A. Narrain, & V. Chandran (Eds.), *Nothing to Fix: Medicalisation of sexual orientation and gender identity* (pp. 223–230). New Delhi: Sage Yoda Press.

Hill, M., Glaser, K., & Harden, J. (1998). A feminist model for ethical decision-making. *Women and Therapy, 21*(3), 101–121.

Indian Association of Clinical Psychologists. (2015). Ethics and code of conduct of clinical psychologists: Guidelines 2012–13. Retrieved April 20, 2015 from http://www.iacp.in/node/159

International Society for Mental Health. (2000). *International Society for Mental Health Online (ISMHO) Principles*. Retrieved December 15, 2013 from http://www.ismho.org/suggestions.asp#Principles

Isaac, R. (2009). Ethics in the practice of clinical psychology. *Indian Journal of Medical Ethics, 6*, 69–74.

Jacob, P., Golhar, T., Seshadri, S., Nandan, R. M., & Purushothaman, K. (2014). Child and adolescent mental health in the juvenile justice system in India: challenges and initiatives. *Adolescent Psychiatry, 4*(4), 278–283.

Kitchener, K. S. (1984). Intuition, critical evaluation and ethical principles: The foundation for ethical decisions in counseling psychology. *Counseling Psychology, 12*, 43–55.

Knapp, S., & VandeCreek, L. (2004). A principle-based analysis of the 2002 American Psychological Association Ethics Code. *Psychotherapy: Theory, Research, Practice, Training, 41*(3), 247.

Kurpad, S. S., Machado, T., & Galgali, R. B. (2010). Is there an elephant in the room? Boundary violations in the doctor-patient relationship in India. *Indian Journal of Medical Ethics, 7*(2), 76–81.

Margolin, G. (1982). Ethical and legal considerations in marital and family therapy. *American Psychologist, 37*(7), 788–801.

McNamee, S. (2009). Postmodern psychotherapeutic ethics: relational responsibility in practice. *Human Systems: The Journal of Therapy, Consultation & Training, 20*, 57–71.

Miller, P. M., Bener, A., Ghuloum, S., Commons, M. L., & Burgut, F. T. (2012). Differences and similarities in cross-cultural perceptions of boundaries: a comparison of results from two studies. *International Journal of Law and Psychiatry, 35*, 398–405.

Nair, M. R. H. (2015). RTI v medical ethics: Some questions arising from the recent decision of the chief information commissioner under the RTI Act. *Indian Journal of Medical Ethics, 12*, 182–185.

Narayan, C. L., & Shekhar, S. (2015). The mental health care bill 2013. A critical appraisal. *Indian Journal of Psychological Medicine, 37*(2), 215–219.

National Academy of Psychology. (2010). *National Academy of Psychology (NAOP), India Ethical Principles for Psychologists*. Retrieved December 15, 2015 from https://docs.google.com/viewer?a=v&pid=sites&srcid=bmFvcGluZGlhLm9yZ3x3d3d8Z3g6MTQ5YTUzNTQyOWMxZjZhNA

National Association of Social Workers. (2008). *Code of ethics of the national association of social workers*. Retrieved from http://www.naswdc.org/pubs/code/default.asp

Orinam. (2014). *Indian Psychiatric Society official statement: Homosexual is not a mental illness*. Retrieved December 20, 2015 from http://orinam.net/indian-psychiatric-society-homosexual-not-mental-illness/

Pomerantz, A., Ross, M. J., Gfeller, J. D., & Hughes, H. (1998). Ethical beliefs of psychotherapists: Scientific findings. *Journal of Contemporary Psychotherapy, 28*, 35–44.

Pope, K. S., & Bajt, T. R. (1988). When laws and values conflict: A dilemma for psychologists. *American Psychologist, 43*, 828–829.
Pope, K. S., Tabachnick, B. G., & Keith-Spiegel, P. (1987). Ethics of practice: The beliefs and behaviors of psychologists as therapists. *American Psychologist, 42*(11), 993–1006.
Pope, K. S., & Vetter, V. A. (1992). Ethical dilemmas encountered by members of the American Psychological Association: A national survey. *American Psychologist, 47*(3), 397–411.
Pope, K. S., & Vasquez, M. J. T. (2016). *Ethics in psychotherapy and counseling: A practical guide* (5th ed.). NJ: John Wiley & Sons Inc.
Proctor, G. M. (2014). *Values and ethics in counselling and psychotherapy*. London, UK: Sage.
Slack, C. M., & Wassenaar, D. R. (1999). Ethical dilemmas of South African clinical psychologists: International comparisons. *European Psychologist, 4*(3), 179.
Sporrong, S., Arnetz, B., Hansson, M. G., Westerholm, P., & Hoglund, A. T. (2007). Developing ethical competence in health care organizations. *Nursing Ethics, 14*(6), 826–837.
Strous, R. D. (2007). Psychiatry during the Nazi era: Ethical lessons for the modern professional. *Annals of General Psychiatry, 6*, 8.
Sullivan, K. (2002). Australian psychologists' ethical beliefs and behaviours. *Australian Psychologist, 37*(2), 135–141.
Tolich, M. (2014). What can Milgram and Zimbardo teach ethics committees and qualitative researchers about minimizing harm? *Research Ethics, 10*(2), 86–96.
Toulmin, S. (1996). Introduction. In S. Toulmin & B. Gustavsen (Eds.), *Beyond theory: Changing organizations through participation* (pp. 1–7). Philadelphia: Benjamins.
Walker, K. D. (1994). Notions of "ethical" among senior educational leaders. *The Alberta Journal of Educational Research, 40*, 21–34.
Zhang, A., Qian, M., & Yao, P. (2007). A survey on ethical beliefs of Chinese counselors and psychotherapists and comparison with American professionals. *Chinese Mental Health Journal, 21*(1), 55–61.
Zhao J., Cheng, W., Fu, S. (2009). A multi-center survey on dual relationships and associated factors among Chinese counselors and psychotherapists. *Chinese Medical Ethics*. Retrieved April 12, 2016 from http://en.cnki.com.cn/Article_en/CJFDTOTAL-XNLX200905014.htm
Zhao, J., Ji, J., Yang, X., Yang, Z., Hou, Y., & Zhang, X. (2011). National survey of ethical practices among Chinese psychotherapists. *Professional Psychology: Research and Practice, 42*(5), 375–381.

Chapter 2
The Ethical Private Practitioner

Rathna Isaac

Abstract Psychotherapy is increasingly being carried out in private practice settings across India. This highlights the need to understand and address ethical concerns, unique to private practitioners. The private practice setting is described as one with greater freedom to act, coupled with greater ethical responsibility. Ethical practices and challenges through the course of therapy from setting up a practice, advertising, informed consent, competence, documentation; through to termination and therapist unavailability are discussed. An attempt has been made to define the scope and limitations of private practice. Concerns around training, qualifications and continued professional development are explored with specific reference to the Indian setting; where creating a uniform standard of care is a special challenge. The business end of therapy and the impact on practitioner, client and therapeutic agendas are explored. A comprehensive and simple model for ethical decision-making is illustrated with an example. Where relevant, suggestions designed to help one build ethics into the structure of one's professional life, have been made. The article is relevant for all mental health practitioners engaging in psychotherapy.

Keywords Private practice · Ethics · Indian setting · Psychotherapy

2.1 Introduction

Being ethical is about more than following a set of guidelines and getting signatures where appropriate. It is about thinking and desiring to act ethically; understanding ethical principles and consciously developing one's personal and professional identity along ethical lines (Anderson and Handelsman 2010). For a mental health professional working in private practice, ethical decision-making is often carried out alone. This paper is intended to assist the private practitioner in understanding, preventing or resolving ethical challenges. It could be useful for psychologists,

---

R. Isaac (✉)
Private Practice, Bengaluru, India
e-mail: rathnaisaac@gmail.com

P. Bhola and A. Raguram (eds.), *Ethical Issues in Counselling and Psychotherapy Practice*, DOI 10.1007/978-981-10-1808-4_2

psychiatrists, psychiatric social workers and counsellors; though it is presented largely through the lens of clinical psychology.

2.2 Working as a Private Practitioner

To illustrate the special challenges faced by private practitioners, I only have to remember what it was like to launch a practice.

I was trained in a well-known tertiary care mental health hospital and training centre in South India, and spent many years cocooned in a safe institutional blanket. I did not need to engage with the client until they had completed all administrative procedures, had been evaluated by a psychiatry consultant and referred for therapy. Clients were less concerned with who 'I' was, than they were with the fact that I was a therapist recommended by the institution. I had no control over the room, its décor, or privacy; neither was I held responsible for them. I did not need to concern myself with billing except for keeping track of the number of sessions. Documentation was regulated as per institutional procedure and the institution took responsibility for maintaining confidentiality and availability of records. Supervision was mandatory and interdisciplinary discussion was the norm. The huge library offered easy access to all the best literature and new ideas in the field which were eagerly discussed both in formal and informal settings.

As a private practitioner, sitting alone in my room, far away from everybody else, I felt truly out in the cold. I was cut off from the support system I had grown used to. Just being a good therapist was suddenly not enough. On the other hand, I had a lot more say in the type of clients I chose to work with, on my timings, on the space that I occupied and in how I chose to proceed with the therapy. Clients came to see 'me' and felt freer to ask about my training and qualifications. They looked around my room and observed and inferred what they might. There was no one looking over my shoulder. No judgement; but no help either.

Private practice offers more freedom and flexibility with work, but also a greater burden of clinical and ethical responsibility (Brennan 2013).

Brennan (2013), in a comprehensive article on ethical issues in private practice identifies four dimensions of ethical knowledge crucial to private practitioners.

1. *Understanding the foundations and principles of ethical conduct*: This includes being aware of various codes of conduct and reflecting on the principles underlying them. For instance, psychologists practicing in India need to be familiar with the ethical codes of the Indian Association of Clinical Psychologists (2015) and the Rehabilitation Council of India (1998). Awareness of the laws of the land as related to the provision of psychological services is also vital as private practitioners typically set up their own system of service delivery. The Mental Health Care Bill (2013) is a good reference in this regard (Narayan and Shekhar 2015).

2. *Understanding the self as the agent of ethical action*, or *introspective ethics:* This refers to increased self-awareness that helps anticipate potential ethical lapses. Brennan (2013) recommends three areas of self-reflection for the private practitioner: personal history (positively or negative emotion laden events/experiences), emotional-temperamental (personality and typical interpersonal stance as well as current stressors) and conventional-functional (the ability to organize and manage yourself, including keeping/scheduling appointments, records, track of fees and completing paperwork).
3. The need to establish an *ethical decision-making process.*
4. *Keeping current and attending to self-care*: The private practitioner needs information about evolving codes and new areas of ethical concern; for instance, the use of text messaging, social media and online interventions. The private practitioner also needs to engage in regular self-care practices, be aware of issues like compassion fatigue and burnout (Figley 2002) and seek suitable and timely help for the same.

Issues like informed consent, confidentiality, minimizing multiple relationships, creating and maintaining boundaries, self-disclosure, prejudice and openness remain ethical guideposts for all mental health professionals, and as such are crucial for private practitioners as well. While ethical concerns and principles themselves are more or less universal, their application may pose particular challenges to the private practitioner. The next few sections will highlight challenges unique to private practice. The following areas will be addressed:

- *Starting a therapeutic relationship* (a) setting up a practice (b) advertising and accessibility and (c) informed consent.
- *Sustaining a therapeutic relationship* (a) competence (b) continued professional development (c) the business side of therapy (d) documentation (e) complex issues (f) ethical decision-making.
- *Ending a therapeutic relationship* (a) refusing clients (b) termination (c) therapist unavailability.

2.3 Starting a Therapeutic Relationship

2.3.1 Setting up a Practice

For the private practitioner, being 'professional' and keeping to time and commitments is entirely between the self and the client. As the individual with more power in the equation it is up to the professional to maintain standards of fidelity and transparency in the day-to-day operations of the practice (Brennan 2013). In addition, ethical considerations can be built into the structure of the practice itself, to minimize the likelihood of violations and dilemmas.

2.3.1.1 Treatment Set-up and Boundaries

Maintaining appropriate boundaries between a therapist and a client is made easier with a more neutral space and personal style, as these minimize the likelihood of inadvertent self-disclosure. A fair number of practitioners work from their homes. Here, the scope for self-disclosure is greater and obviously harder to control (Zur 2011). A client coming into your home may feel a bit more like a guest and they may ask more personal questions about their surroundings. Not all personal queries are intrusive. For instance, clients from a different socioeconomic status may feel that the therapist will not be able to grasp their worldview. The author recommends non-defensive but clear boundary setting.

A statement along the lines of "I am happy to respond to any questions that you feel are relevant to your view of me and our professional relationship; however I would appreciate it if we kept away from discussing my personal life, as that may distract from the purpose of therapy" could be useful.

This statement recognizes that the principle underlying less self-disclosure is to keep the therapy space neutral and non-judgemental, and not so much to keep the client 'away' from the therapist (Roberts 2012). It demonstrates how understanding basic principles can help you frame positive statements for clients. It models awareness of own needs (in this example, the need for privacy), and willingness to be constructively assertive about them.

2.3.1.2 Advertising and Accessibility

The private practitioner needs to be accessible to potential clients. Appropriate contact procedures and information can be provided, either on a website or visiting card. If available on the phone, it is useful to specify how (through text/call/email) and when (when you will receive calls, expected turnover time for texts or emails) clients can contact you. On a website, one needs to limit information provided to the description of qualifications, area(s) of interest, fees, availability, address and contact information (Tran-Lien 2012). All information on the website or in any form of advertising needs to be accurate and updated when applicable (American Psychological Association 2010). False advertising and tall claims call the integrity of both the professional and the profession into question. Creating unrealistically high expectations can be actively harmful to clients.

2.3.1.3 Informed Consent

As with any relationship, the first impression sets the tone for subsequent contact and a good intake session is half the battle won. In the author's experience, private practice clients tend to be more aware and articulate about their preferences. They need to be told about the expected process of therapy, their role, the therapist's role, estimated frequency/duration (and by implication, cost) of treatment, likelihood of success, etc. Clients also need to be made aware of alternate treatment options, both psychological and non-psychological. Their understanding of this material forms the basis for informed consent (Bearhs and Gutheil 2001). Some practitioners find a written informed consent form to be useful, particularly in the private practice context. Spelling out the contract can leave both client and therapist feelings safer. The author has evolved a brief sample form in a checklist format (see Appendix B). This format offers both flexibility and simplicity; it encourages discussion and is not too long. Both Bearhs and Gutheil (2001) and Brennan (2013) emphasize that informed consent is an ongoing process. As therapy evolves, new areas may come up for discussion. For example, if a family member is coming in for a consultation session in an individual therapy, confidentiality and boundaries will need to be redrawn.

Along with describing the process of therapy and discussing items on the informed consent form, the author also recommends being explicit about details surrounding the process of scheduling and rescheduling appointments, extra session contact, contact on social media and fee structure.

2.3.2 Sustaining the Therapeutic Relationship

2.3.2.1 Competence

Ethical codes emphasize that one should not do anything one is not trained/qualified for. Professionals should only use only those forms of treatment they have received training in, for clients they are qualified to see (British Psychological Society 2009). This seems pretty straightforward, but it is a fairly tall order for the Indian private practitioner. In a survey of 250 mental health practitioners across the country, lack of competence was listed as the single biggest limitation, experienced by 41 % of the sample (Bhola et al. 2012).

The issue here is that it is hard to decide what exactly one is qualified to do. While all mental health professionals have *some* training; there are various levels of training (Misra and Rizvi 2012). Mental health professionals include clinical psychologists with PhDs or MPhils, psychiatrists with MDs or DPMs, psychiatric social workers with PhDs or MPhils, masters level graduates in psychology, counselling or social work, counsellors who have undergone accredited or 'not-so-accredited' counselling courses, etc. The training system in India typically offers some insights into all the major schools of therapy, without a very in-depth training in any one school. We do not usually have access to treatment manuals. For instance, can one say they are doing dialectical behaviour therapy (DBT; Linehan 2014) when they have not done Linehan's course? What if one does not even have the manual? Should one then refuse to see clients who require DBT? And if one does refuse to see such clients, who will? The APA code (American Psychological Association 2010) says that if there is no qualified person to see a client, the next best or closest in qualification may do so, as long as they commit to training themselves as much as possible. But what does that actually translate into here?

It is up to the practitioner to be aware of what they are qualified to do and to communicate their scope and limitations clearly to the client. Private practitioners may feel more pressure to try out a new treatment. This pressure often comes from clients themselves and many of us may end up attempting work that we are not fully trained to do. One must be willing to draw boundaries and refuse to undertake certain treatments if not adequately trained, as illustrated in Box 2.1.

Box 2.1 Drawing Boundaries*
A long-term individual therapy client was very upset that I would not do Eye Movement Desensitization and Reprocessing (EMDR) with her. She had symptoms of post-traumatic stress, and had read that this was the best form of treatment for her condition. She was unwilling to see anyone else this late into therapy, and expressed every confidence that if I read a bit, I could do it with her. I needed to explain that it was a technical procedure that required special training that I did not have. I explained the process I would use to address her symptoms and the theories on which my choices were based. She agreed only reluctantly, and did feel short-changed and a bit let down by my refusal to comply with her request. I still felt it would be inappropriate for me to try and execute a treatment which I was not sure of.

**Note* All case illustrations in this chapter are composites and do not refer to specific clients.

2.3.2.2 Continued Professional Development

The ground reality of mental health in India is a huge gap between numbers of qualified mental health professionals and actual need for the same (Isaac 2009; Murthy 2011). Given this gap, the author recommends that every effort be made to improve on ones' training through focus on continued professional development. We do not have the luxury of sticking to our comfort zones. It is extremely important to expand our knowledge with reading and supervision. We may *need* to try out newer forms of treatment or get acquainted with new client groups; *but* we should always do so under direct or online supervision. The author recommends the following practices to help identify strong points and comfort zones and to keep current with new methods (see Box 2.2).

Putting time and effort into continued professional development can give both financial rewards and deeper professional satisfaction.

Box 2.2 Good Practices for Continued Professional Development

1. Consider the syllabus that has been covered during training, in terms of diagnostic groups, forms/schools of therapy, child or adult client populations; to determine what you know and the potential scope of your practice.
2. Consider areas of comfort or expertise created or enhanced by supervision opportunities or because of interest and available reading material.
3. Consider special areas of discomfort or poor training. Seek additional resources to address these gaps if possible.
4. Work together to organize continuing education programmes to help address new issues (for e.g., road rage)/client groups (for e.g., children of divorce)/diagnostic categories (for e.g., eating disorders).
5. Do not overstate the scope of your practice/training. For instance, a weekend workshop on couple therapy does not qualify you to start seeing couples. If you do choose to engage in a new area or if clients are unable to access more qualified professionals (which necessitates that you see them), do inform clients of your limitations and extra measures you are taking to ensure that they receive quality help (Brennan 2013).
6. Try to do case-based reading from classic textbooks as well as currently available online information.
7. Always have a supervisor/someone you respect who you can discuss cases with. This can be a formal arrangement where you pay for supervision or an informal arrangement, where a group of psychologists meet and discuss therapy (peer supervision).
8. Always have a therapy plan and frequently review the same.
9. Engage in self-reflection. You can use the areas suggested by Brennan (2013) as a good starting point.
10. Keep session notes and keep time for reflection on them.
11. Listen to feedback from your client.
12. Be willing to accept when you are out of your depth. Examine whether your desire to refer is a competence issue or a transference issue.
13. Refer to other disciplines where necessary. For instance do not start sex therapy without a review by a medical doctor.
14. Become a member of a society, attend conferences and CMEs, talk to colleagues and find out what standard practice is. Be willing to share about your practice.

2.3.2.3 The 'Business' of Therapy

Both clinicians and clients can have difficulties with the business end of the relationship. Reconciling the unconditional acceptance and support of the therapeutic relationship with the mundane reality of "This is how much I charge" is not very easy. When the same person is playing both roles, the therapeutic contract and the business contract can impact each other, as illustrated in Box 2.3.

Box 2.3 The Business of Therapy
A young client (21 years) once wanted to negotiate a small reduction in fee. She was aware that I worked on a sliding scale. My agreeing or disagreeing would have more of an impact on our therapeutic relationship than on my financial condition. Agreeing would have set a bad precedent of inconsistent rules and porous boundaries, and violated the ethical principle of 'justice'. However, I could see that for her, the meaning of the interaction was more to do with personal power and a belief that the world could not be trusted to recognize/acknowledge/help her in any way. I needed to decide which course of action would cause the least harm to her, while preserving my professional integrity and creating a sustainable therapeutic contract.

Private practitioners are usually left to set their own fees. Determining how much an hour of one's time is worth can be a nerve wracking experience indeed. The Canadian Psychological Association code (CPA 2000) recommends setting fees that are "fair in light of the time, energy, and knowledge of the psychologist and any associates or employees, and in light of the market value of the product or service" (p. 10). But how does one decide what is fair? A consensus or at least communication among private practitioners with similar qualifications practicing around the same area will be very useful in this regard.

If using a *sliding scale* of payment, Brennan (2013), asks practitioners to consider: (a) Are levels on the sliding scale fixed? (b) Are income and other criteria to access lower fees clear and transparent? (c) Are all clients aware of the existence of a sliding scale? (d) Does it accommodate to clients whose income may change during the course of therapy so that they move up or down the scale? (d) Are clients fully informed about billing practices? (e) Do they pay at the beginning or the end of the appointment, and who do they pay? A written document spelling out billing practices can help increase clarity and consistency.

Many practitioners charge a fee for missed sessions known as a *cancellation fee*. It is usually applied only for last-minute cancellation of a session. It may be the full session amount or a percentage of the same. A cancellation fee helps the client to take therapy seriously and also protects the income of the practitioner. The cancellation fee is of particular relevance to private practice, as each hour contributes to overall monthly income and cancelled sessions can lead to a dent in income and possible negative feelings toward the client.

The provision of mental health services by a professional is recognized as a paid activity, and we need to be comfortable with the fact that we do this for money. It is important, however, not to take advantage of the trust or dependency of the client to force services on them; for example, in recommending an assessment or further sessions of therapy. Referrals should be made and received on the basis of the best interest of the client and not for monetary gain. The practice of taking a percentage or a 'cut' off referrals made is clearly and unequivocally unethical.

Finally, some clients may not be able to afford on going or very long-term therapy. The practitioner can refer such clients to another centre/therapist that charges less, offer to continue at a lower slab on the sliding scale, or even accept services as barter. However, if the client does not wish for any of these solutions, and does not pay their fees, the therapeutic contract and ethical liability are terminated (American Psychological Association 2010).

Third party payments The issue of "who is the client" or "whose needs the therapy should address" is particularly relevant when parents are paying for sessions for their (legally) adult children, spouses or parents. The person who is making the payment may assume that the therapeutic contract is being made with them and may often request the practitioner push for a particular change. They may ask the therapist to "make the client realize…" or "make her stop…" something. This is a tricky situation as the practitioner can neither ignore their perspective nor give into it completely. The following suggestions are partly based on recommendations by de Sousa (2010), a child psychologist, on dealing with parents and families in the treatment of children. They are relevant for adults as well, and reflect our interpenetrative culture.

- Discuss the validity of the expectation/suggested agenda with the client. Not all expectations are harmful/negative. Perhaps the client also wants to make similar changes in themselves.
- If the client feels that the expectation is unfair, but is disempowered to negotiate with their family member, help them learn how they can do this. If they need extra support, you can offer them session time to facilitate this conversation.
- If the family member requires psycho-education to understand the limits and potential of the client, it is the duty of the psychologist to try and provide this information.
- Do not negotiate on behalf of the client, but empower them to negotiate for themselves.
- Refer a client to family therapy if issues seem very intractable.

2.3.2.4 Documentation

'Documentation' is a term that raises some dread in the hearts of most practitioners. While we all know we need to document, how much and what is documented varies from practitioner to practitioner. In the absence of institutional guidelines for the same, each private practitioner tends to evolve their own parameters.

In the context of psychotherapy, documentation ideally covers the following areas:

- Name and contact information of the client(s)
- Informed consent, either in the context of a written form or notes on the content discussed
- Presenting problem, reason for referral and source of referral
- Case work up/history
- Medication and diagnosis if applicable (both psychiatric and non-psychiatric)
- Therapy plan
- Session notes including date, time, aim, content, process and observations
- Details of extra session contact
- Changes in the therapy plan and reasoning for the same
- Ethical dilemmas and their resolution
- Termination notes

Very few people actually enjoy tedious paperwork. Detailed documentation is often the first thing to fly out of the window as we gain our feet and confidence in the professional world. Not only are there limitations of time, we realize that we can go from session to session without the notes—and clients do not really notice or ask about them. Adequate documentation, however, is crucial for competent practice.

Let us apply Brennan's (2013) perspective to the issue of adequate documentation: Many practitioners see documentation as an extra burden, required to protect oneself legally. It is important to understand that good documentation reflects the principles of openness and integrity and enables you to confirm that what you are doing with the client is beneficial and not harmful. The process of documentation aids self-reflection. It is often during the writing of the session notes that we become aware of our own reactions and feelings; it is only after the session that we have time to connect session experiences to our personal lives. The process of documentation enables ethical decision-making and helps us keep track of evolving ethical issues/dilemmas. Finally, going through our notes can help to pick up larger themes of commonly faced issues (text messaging with many clients) or need for self-care (a therapist might notice that they feel emotionally overwhelmed with many clients and not just one).

If you deepen your understanding of why you document and what to document, it gives meaning, purpose and relevance to your effort. This underlines the perspective that being ethical is an ongoing process that permeates all aspects of our professional lives.

Clients need to be informed that session notes would be made, and also what is being recorded. If the client reveals information that they do not wish to have documented, it is required for one to leave it out of written records, unless it is central to the understanding of the client and therapy-related decisions. Even where it needs to be recorded, the information should be retained in the most innocuous form possible. Avoid conjecture and when you are stating an opinion, clearly state it (Canadian Psychological Association 2000). For example, if the client brings up sensitive material that does not have a direct impact on the issue (for example, my father was accused of bribery and corruption by his employers, when I was ten years old), it need not be recorded. If the material is relevant (…and therefore he lost his job and we had to shift to another city…), it can be recorded as 'Client moved to B… when he was ten due to work difficulties faced by his father'.

Complete documentation can prevent ethical violations from arising as it encourages reflection and thoughtfulness about therapeutic choices. It may also be a private practitioner's only defence against claims of malpractice. Good documentation serves both the practitioner and the client.

Storage of documents Having written or typed out the material, the practitioner also needs to be sensitive to how it is stored, and who has access to it. If other staff is being employed, for example, in a secretarial role, their access to records, sessions and clients need to be clarified. They would need to understand the necessity for confidentiality and discretion. Physical papers need to be stored in a locked space, which others do not have access to. Digital material should be stored in a password-protected and encrypted form. If you are storing contact information on your phone, it may be better to have a separate work phone so that this information is not accessible to others.

Use of client data If the practitioner intends to use client data for teaching, supervision, or other professional activities, they need to get permission from the client for this. If this is a regular practice for the practitioner, it is a good idea to include a sentence about this in the informed consent form itself. If information is going to be used in research or publication, it may be better to get separate written permission for the same. The practitioner needs to be aware of inherent power differentials between themselves and the client and make every effort to ensure that the client understands that this is completely their choice and that refusing to allow their data to be used will not have an impact on the therapeutic relationship.

2.3.2.5 Scope of Private Practice

Private practitioners can handle most of the issues that clients may bring in. Developing a strong network of psychiatrists, psychologists and social workers can help you address more complex issues with multiple ethical, legal and competence-based questions. However, some clients may do better with institutional support from a multidisciplinary team. In the author's experience, actively suicidal clients; clients with interrelated medical, neurological, psychiatric and psychological problems (for instance, severe eating disorders); clients requiring frequent admission (for instance poly-substance dependence), or clients where the entire family needs help and requires individual and family therapists to work in tandem with each other and with psychiatrists, are best treated in an institutional set up.

When seeing a more complex case, particularly suicide, it is doubly important to: (a) explain the limits of confidentiality; (b) get an emergency support contact number; (c) establish procedure for such emergency contact at the onset of therapy itself; (d) if making referrals, identify the other members of the 'team'; (e) and explain how and when information will be shared with them (Hawgood 2015). We may feel reluctant to bring these issues up with highly emotional clients, especially when we are trying hard to create a rapport. But failure to do so in the beginning can lead to clients feeling unsupported or 'cheated' and lose trust in the therapeutic process entirely.

Psychological services are increasingly coming under the legal scanner. Appropriate documentation (as described) can ensure and explain clinical and ethical choices. Clients may also come up with legal queries, most often related to divorce, alimony, child custody, child protection, etc. They may ask when they should time their divorce, who should file, the laws under each religion/marriage act, the amount of alimony they can expect to give/receive, etc. Clients may also ask about their legal recourse in cases of abuse (especially child abuse) or other wrongful behaviour. While it is useful for a practitioner to be aware of the laws relating to mental health issues, it is important to refer them to proper legal counsel as well. "But you said he wouldn't get custody, and I trusted you" is not a phrase you want to hear in session.

Narayan and Shikha (2013) reviewed the legislation related to marriage and divorce in the context of different religions in India; for example, Hindu Marriage Act (1955), Special Marriage Act (1954), Muslim Marriage Act (1939), Indian Divorce Act (1939), Indian Divorce Act (Amended 2001). Some of the other relevant laws and acts include the Indian Succession Act (1925), guidelines from the National Commission for the Protection of Child Rights (NCPCR), Protection of Women from Violence Act (2005) and Protection of Children Against Section Offences Act (2012).

2.3.2.6 Ethical Decision-Making: When Principles Conflict

A conscientious practitioner may be able to avoid most ethical violations. However, there are situations where any choice has ethical implications, and no choice is completely right. Having a process of how to think through such ethical dilemmas decreases the likelihood of an ethical blunder. This process is referred to as ethical decision-making and will be illustrated using a model based on the ACA model (American Counselling Association) as described in Brennan (2013) and the Canadian Psychological Association models (CPA 2000).

Box 2.4
A couple (Mr. A and Ms. B) came for couples' therapy, following disclosure that the husband had had an affair. The couple was doing well, had re-established trust and was working on building their relationship. At this stage, through an inadvertent disclosure by the husband. I became aware that his affair partner (Ms. C) was also an ongoing client.

For a situation as in Box 2.4, the American Counselling Association (ACA 2005) recommends first determining the nature and dimensions of the dilemma. Issues arose at two levels: (a) What to tell the clients and what to keep confidential. While principles of integrity and openness required that I revealed information that clients would find pertinent, it also seemed that it would cause harm to all three individuals as well as the couple relationship. It would be particularly damaging to Ms. B, to have to share both her husband and her therapist with Ms. C. Ms. C might perceive it as a betrayal that the therapist was working on building a relationship that she hoped would break. Mr. A might find his ambiguities emerging and his loyalty being questioned again! (b) To examine if this would interfere with carrying out the therapy work effectively. It is difficult to keep such knowledge out of one's consciousness. In couple sessions, there would be a lot more information about one person (Mr. A) than the other. This could lead to a skew either towards or away from him. I would be aware of aspects of Ms. C's situation that Ms. C herself had not shared with me. How then is one to have a genuine and open therapeutic relationship?

The Canadian Psychological Association (2000) describes a comprehensive ten-step process for ethical decision-making.

1. *Identify the stakeholders or individuals/groups likely to be affected by the decision* There were three stakeholders involved (Mr. A, Ms. B and Ms. C), all of who were responding well to therapy. The integrity of the therapist and that of the profession could also be affected by the choice.
2. *Identify ethically relevant issues and practices based on both the client group and the treatment setting* This was an issue of multiple relationships, couple and individual clients, being seen in a private practice. The therapist was the only mental health professional involved with all three clients. Questions arose around principles of integrity, respect for client autonomy and rights, and potential harm to the clients.
3. *Consider how personal bias/stress might influence your choices* Not telling the clients would create personal discomfort, and the influences of the dual relationship might remain largely underground. Telling them on the other hand, would definitely be a 'difficult' conversation to have, and challenge my skills and rapport. It would also open the door for clients to question my values. The appropriate choice may be influenced by how well I am able to contain the dual relationships and remain genuine with all three parties. My values surrounding marriage and extramarital affairs (who is the victim? who needs protection?), surrounding openness to self and appropriate resolution of emotions (emotion focused versus pragmatic, self-exploration versus reducing negative emotions, stability versus exploration) will all have an impact on the final decision and need to be acknowledged.
4. *Develop several courses of action* (a) I could keep the dilemma to myself; handle the cognitive and emotional dissonance through supervision, a clear therapy plan and self-reflection; and allow both therapies to continue to termination. (b) I could disclose the dual relationship and enable clients to make an informed decision about what to do.

5. *Analyse short-term, ongoing and long-term risks and benefits for all stakeholders including professional bodies and the profession itself* Potential consequences of choice 'a' include: the burden of the 'secret' could become too much to bear and intrude on the therapy process or clients find out about the dual relationship. Mr. A and Ms. C may be aware of the dual relationship all along and assume that I am colluding with them. Choice 'b' could damage progress made at both individual and couple levels. The couple/Ms. C might discontinue therapy, therapist may be asked to *choose* which client she will continue with, etc. On the positive side, it could also create a space for greater integration, forgiveness and understanding for all three and ultimately benefit clients and therapist alike.
6. *Choose a course of action after applying existing ethical standards and values* Let us assume that (under supervision) I chose not to disclose the dual relationship as this would be deemed too harmful for all three clients whose stability was still fragile. I might conclude that the burden of the knowledge should rest with the professional and not the clients, and that non-maleficence was more important than integrity. I may be influenced by other factors like unavailability of alternate therapists to refer to or reluctance to lose income from three clients at the same time.
7. *Take action and take responsibility for the consequences* Once the decision is taken, I need to remain aware of potential pitfalls; and monitor my neutrality and objectivity, as well as client comfort and progress in therapy.
8. *Evaluate the results* I might find that I could not keep information out of clinical responses and decisions. I would then need to consider the alternate choice.
9. *Correct any negative consequences and re-engage in the decision-making process if issue is not resolved* Subsequently, I might choose to reveal the dual relationships, as I observed that it was not possible for me to remain neutral and provide the best quality care for all three clients. I may conclude that such a secret would not be easy to keep, as all three clients do know each other. Also, that my integrity would be completely damaged by keeping further secrets rather than modelling openness and willingness to confront and address difficult emotions and situations. Ultimately, this could damage all three clients beyond my ability to help them. It would be easier to make this choice if I felt that the clients have progressed far enough in therapy and that relationship decisions have already been made, therefore it will be easier for clients to handle the dual relationships successfully.
10. *Take action to prevent future recurrence of this dilemma* This could include discussing such possibilities during initial sessions, developing a more effective network of referrals and/or clarifying and crystallizing own personal and professional values.

As illustrated, making an ethical decision is not easy. There are circumstances in which ethical principles *do* conflict with each other (beneficence/non-maleficence versus respect/integrity in the example above). In general, it is recommended that non-maleficence be the governing principle, while relative weight to the other principles varies from case to case. Such a decision can only be taken contextually. Using a written format ensures that each situation gets the time and attention it requires.

2.3.3 Ending the Therapeutic Relationship

Just as relationships begin, so too do they end. Sometimes they stop before they get started, sometimes they end messily or prematurely and sometimes they end well.

2.3.3.1 Can You Refuse a Client?

It is unethical to refuse a client on an arbitrary or discriminatory basis. Kaplan (2014) quotes the American Counselling Association code of ethics (ACA 2005), to emphasise this point. 'Counselors do not condone or engage in discrimination based on age, culture, disability, ethnicity, race, religion/spirituality, gender, gender identity, sexual orientation, marital status/partnership, language preference, socioeconomic status, or any basis proscribed by law' (p. 10).

However, this does not mean that one needs to see all clients who call to fix an appointment. Kaplan (2014) distinguishes between 'inability' to see a client and 'unwillingness' to do so. Time constraints, inadequate competence to deal with an issue and harmful multiple relationships are valid reasons to refuse a client. Sometimes, at the end of the initial session, the practitioner may feel that therapy would be potentially harmful at the time, with no balancing benefits. The pros and cons of therapy need to be discussed and alternate therapists/means suggested.

It could happen that the practitioner cannot take on new clients at that moment because their slots are full. If this happens, the author recommends giving the client an approximate session date and offering to refer to an alternate professional if they wish to see someone sooner. Clients really appreciate our consideration for their time, money and emotional states.

Even where it is permissible or beneficial not to see that particular client, the practitioner still has responsibility for appropriate referral (Kaplan 2014). If the client is unwilling for referral, one can discontinue the relationship with no further ethical obligation (American Mental Health Counsellors Association 2010).

Busy private practitioners may use a waitlist system, where clients will be taken as and when free slots are available. This system is useful in protecting the income of the practitioner. It is one way of handling high case-loads, particularly when there are fewer alternative professionals to refer to. However, the wait may be indefinite and the client might need help immediately. The practitioner must always offer a referral as an alternate to the waitlist.

2.3.3.2 Termination

The duration of therapy is usually based on experience and standard practice more than on manual or RCT-based recommendations (Goldfried and Wolfe 1998). Each practitioner may have a duration that is typical for him or her. For example, the author finds that 10–15 weekly sessions are usual, with a small percentage of clients continuing on for long-term therapy, and of course a small percentage terminating earlier. Clients need to be informed about the average or expected duration, as well as systems of review of therapy to decide on future directions.

Premature termination Sometimes, clients may wish to discontinue therapy before the practitioner feels that they are ready to do so. Respect the right of persons to discontinue therapy at any time. It helps to discuss and validate possible reasons for termination (health, financial, move, therapy not helping). It is important to give clients an understanding of what they need to do if they wish to stop sessions, as this reduces the likelihood of unexplained dropouts. It is also necessary to let them know what to do if they change their minds and wish to restart therapy. Clarify specially about availability; will they need to go back on the waitlist or will they be seen immediately.

Practitioners are also advised to terminate professional services when clients do not appear to be deriving benefit and are unlikely to do so. If the practitioner feels that they have made their best efforts with reading and supervision and are unable to help the client, it is best to refer to someone else or discuss termination. When doing this, it is important not to convey to the client that they are a 'hopeless case' or that their therapist has given up on them. This can be construed as 'abandonment'. Finally, never terminate without a plan in place for further contact as and when required.

2.3.3.3 Therapist Unavailability

This is an issue of particular relevance for the private practitioner, who often has sole responsibility for the well-being of the client. Practitioners may sometimes become unavailable during the course of therapy, due to anticipated or unanticipated events in their personal life. However, they retain responsibility for client care. In such a situation the following are recommended:

- If it can be anticipated (for instance, pregnancy, moving to another city), the practitioner needs to inform clients in advance so they can both plan how to respond. Clients might feel abandoned by therapists experiencing life events. A long-term client was visibly upset at hearing of my second pregnancy. Attending to her feelings about this helped us to begin talking about her own pregnancy experiences and the lack of support she felt around them.
- If the client is moving on to another practitioner, one must do whatever possible to make the transition smooth. Having a joint session with the new practitioner can be very useful in this regard. One needs to remain available to the client and liaise with the new professional during the transition phase.

- If experiencing burnout or other psychological issues, the private practitioner needs to get appropriate help. High caseloads, high levels of trauma in clients, or many clients with similar issues; put practitioners at greater risk (Figley 2002).
- The practitioner should practice self-care activities that help avoid such situations from arising. These could include personal therapy, yoga/meditation, supervision or self-reflection among others (Baruch 2004).
- If the practitioner becomes physically incapacitated, or dies, they need a 'professional will' in place. This is a document that will help clients deal with the loss, and transition as smoothly as possible to new modalities of care. It includes information on accessing/protecting records, notifying clients and ideally, appointing an 'executor' who can meet clients and organize referrals (ACA 2005).

2.4 Conclusion

As Brennan (2013) pointed out, the private practitioner needs to police herself, as there is no one else to do so. Awareness of ethical issues and potential pitfalls is essential. At the same time, one cannot be an effective therapist in an atmosphere of fear and self-protection. When the supporting structure of one's practice is constructed on ethical grounds, the whole edifice is more secure. Both therapist and client can then be freed to play.

References

American Counseling Association. (2005). *ACA code of ethics.* Retrieved from http://www.counseling.org/Resources/aca-codeof-ethics.pdf

American Mental Health Counselors Association. (2010). AMHCA code of ethics. Retrieved from http://www.amhca.org/?page=codeofethics.

American Psychological Association. (2010). APA ethical principles of psychologists and code of conduct. Retrieved from www.apa.org. Ethics Office.

Anderson, S. K., & Handelsman, M. M. (2010). *Ethics for psychotherapists and counselors: A proactive approach.* New York, NY: Wiley.

Baruch, V. (2004). Self-care for therapists: Prevention of compassion fatigue and burnout. *Psychotherapy in Australia, 10*(4), 64–68.

Beahrs, J. O., & Gutheil, T. G. (2001). Informed consent in psychotherapy. *American Journal of Psychiatry, 158*, 4–10.

Bhola, P., Kumaria, S., & Orlinsky, D. E. (2012). Looking within: Self-perceived professional strengths and limitations of psychotherapists in India. *Asia Pacific Journal of Counselling and Psychotherapy,*. doi:10.1080/21507686.2012.703957.

Brennan, C. (2013). Ensuring ethical practice: Guidelines for mental health counselors in private practice. *Journal of Mental Health Counselling, 35*(3), 245–261.

British Psychological Society. (2009). Code of ethics and conduct. Retrieved from www.bps.org.uk/system/files/.../code_of_ethics_and_conduct.pdf

Canadian Psychological Association. (2000). *Canadian code of ethics for psychologists* (3rd ed.). Retrieved from www.cpa.ca

De Sousa, A. (2010). Ethical issues in child and adolescent psychotherapy: A clinical review. *Indian Journal of Medical Ethics, 7*(3). ISSN:0975–5691.

Figley, C. R. (2002). Compassion fatigue: Psychotherapists' chronic lack of self-care. *Journal of Clinical Psychology, 58*(11), 1433–1441.

Goldfried, M. R., & Wolfe, B. E. (1998). Toward a more clinically valid approach to therapy research. *Journal of Consulting and Clinical Psychology, 66*, 143–150.

Hawgood, J. (2015). Working with suicidal clients: Impacts on psychologists and the need for self-care. *InPsych*. Retrieved Sept 14, 2015 from www.psychology.org.au

Indian Association of Clinical Psychologists. (2015). Ethics and code of conduct of clinical psychologists: Guidelines 2012–13. Retrieved April 20, 2015 from http://www.iacp.in/node/159

Isaac, R. (2009). Ethics in the practice of clinical psychology. *Indian Journal of Medical Ethics, 6*, 69–74.

Kaplan, D. M. (2014). Ethical implications of a critical legal case for the counseling profession: *Ward v Wilbanks. Journal of Counseling & Development, 92*, 142–146.

Linehan, M. M. (2014). *DBT skills training manual* (2nd ed.). New York, NY: Guilford Press.

Ministry of Women and Child Development. (2012). The protection of children against sexual offences bill. Retrieved from http://wcd.nic.in/sites/default/files/childprotection31072012.pdf

Misra, R. K., & Rizvi, S. H. (2012). Clinical psychology in India: a meta-analytic review. *International Journal of Psychological Studies, 4* (4). http://dx.doi.org/10.5339/ijps.v4n4p18

Murthy, R. S. (2011). Mental health initiatives in India (1947–2010). *National Medical Journal India, 24*, 98–107.

Narayan, C. L., & Shekhar, S. (2015). The Mental Health Care Bill 2013: A Critical Appraisal. *Indian Journal of Psychological Medicine 37*(2), 215–219. doi:10.4103/0253-7176.155634

Narayan, C. L., & Shikha, D. (2013). Indian legal system and mental health. *Indian Journal of Psychiatry, 55*(2). doi:10.4103/0019-5545.105521

Rehabilitation Council of India (1998). Notification [File No. 5-62/93-RCI]. The Gazette of India: Extraordinary, Part III, Section 4. Retrieved November 13, 2015 from http://www.rci-online.org/ethics

Roberts J. (2012). Think before you get personal. *Psychotherapy networker*. Retrieved December 3, 2013 from www.psychotherapynetworker.org/magazine/currentissue/item/1741-therapist-self-disclosure

The Hindu Marriage Act 1955, Act 25 of 1955. (1955) Retrieved March 23, 2016 from http://highcourtchd.gov.in/hclscc/subpages/pdf_files/4.pdf

The Protection of Women from Violence Act 2005, No 43 of 2005. (2005). Retrieved March 23, 2016 from http://ncw.nic.in/acts/TheProtectionofWomenfromDomesticViolenceAct2005.pdf

Tran-Lien, A. (2012). E-mailing your client: Legal and ethical implications. *The Therapist*, 5/6. Retrieved from https://www.camft.org/

Zur, O. (2011). Self-disclosure and transparency in psychotherapy and counselling: To disclose or not to disclose, this is the question. Retrieved from http://www.zurinstitute.com/selfdisclosure1.html

Chapter 3
Ethics in Child Psychotherapy: A Practitioner's Perspective

Sowmya Bhaskaran T.S. and Shekhar P. Seshadri

Abstract Children represent one of the most vulnerable populations and it is the responsibility of the mental health professional to identify ethical issues and incorporate ethical principles into the practice of psychotherapy. In this chapter, basic ethical principles such as boundaries, privacy, confidentiality, autonomy and competence as they apply to psychotherapy for children and adolescents are discussed. Approaches to challenging situations such as counter transference, child custody disputes, revelation of abuse, multiple relationships and multiple therapists are highlighted. Commonly encountered issues such as online practices, public encounters and termination are reviewed. The chapter uses vignettes as a method to exemplify ethical challenges and each vignette is followed by a discussion of the process of assessing the situation, taking into account all the relevant factors and principles to come to a conclusion regarding the course of action in a given therapeutic situation. A balance of unwavering commitment to their safety, dedication to act in their best interests and perusal of issues from a rights-based perspective is the cornerstone of the ethical practice of child and adolescent therapy.

Keywords Child · Adolescent · Ethics · Psychotherapy · Therapy

Therapy in itself presents an inherent dilemma. The idea that a total stranger can help ease another's pain or that interacting with someone can lead to improvement in one's life is a difficult one to comprehend.

Children are mostly brought to the clinic by their parents, at times against their will because parents are genuinely concerned. They may have been referred from school or college because they have been causing trouble to others. Since they are

S. Bhaskaran (✉)
Insight Clinic, Coimbatore 641004, Tamil Nadu, India
e-mail: bsowmya1984@hotmail.com

S.P. Seshadri
Department of Child and Adolescent Psychiatry, National Institute of Mental Health and Neuro Sciences, Bengaluru 560029, India
e-mail: shekhar@nimhans.ac.in

P. Bhola and A. Raguram (eds.), *Ethical Issues in Counselling and Psychotherapy Practice*, DOI 10.1007/978-981-10-1808-4_3

mostly not self-referred, therapy with children and adolescents requires a partnership which is unlike that of a therapist and an adult patient.

Work with children involves an understanding of developmental issues, the varied nature of presentation of disorders, appropriate management strategies and ethical principles. Thus a practitioner has to integrate the needs of the child, the family's objectives in treatment and the developmental level of the child. One must also keep the sociocultural context that the child is reared in mind while providing care to the child. Their relatively higher dependence and lower autonomy present complex ethical challenges in clinical care (Belitz and Bailey 2009).

The responsibility to uphold ethical standards and incorporate ethical principles into practice lies with the psychotherapist. Because of this responsibility, identifying ethical issues as and when they arise and understanding how one's personal values and beliefs may affect the therapeutic relationship, are paramount to maintaining professionalism and ethics in psychotherapy (Roberts et al. 2002).

When a child is taken up for therapy, it is the therapist's obligation to create and protect the integrity of the psychotherapeutic space. Boundaries, privacy, confidentiality and autonomy are components of this space (Ascherman and Rubin 2008).

3.1 Basic Principles

3.1.1 Respect for Boundaries

Children and adolescents need a more personal and revealing mode of communication. Their growth in therapy is mostly through the relationship experience. Therefore, understanding boundaries and boundary issues becomes vital to maintain a consistent therapeutic relationship with young people.

Although some boundaries are firm, such as avoiding sexual contact with any current or former patient (American Psychiatric Association 2001), it is now recognized that some boundaries, especially with children and adolescents are more fluid.

Children often ask about the therapist's age, marital status, whether they have children and may also give invitations to visit their homes. Adolescents may ask about personal experiences like substance use or views on controversial subjects. Such questions indicate curiosity and provide opportunities to understand the young person. Motivations for these questions are elucidated without necessarily answering these questions. Being 'neutral' to these questions means that we neither encourage nor condemn them, but remain interested. Here it is important to gauge the meaning behind the question. Keeping the therapeutic need and developmental level in mind, a response that is congruent to the child's intent behind the questions is given. This construct of "therapeutic neutrality" helps the therapist secure a position that protects the therapeutic space for the child. The therapist invites discussion rather than expression through action which confuses boundaries. However, it is not necessary that the therapist always remains a 'blank screen' because he is inevitably a participant/observer in the therapeutic dyad (Sullivan 1940).

Children and adolescents tend to think and communicate in concrete terms. Communication at a personal level, which inevitably involves some self-disclosure, is more effective. Children and adolescents are less able to understand that the therapist is playing a role; governed by certain rules (Gaines 1995). They experience the relationship as real, and the therapist's expressions or lack of expressions as his or her true feelings. In this context, a conversation without some therapist self-disclosure feels cold and rejecting (Gaines 2003). Tactful and carefully planned therapist self-disclosure can be a natural part of therapeutic interaction and help in fostering engagement with the child.

3.1.2 Respect for Autonomy

Adults are defined as individuals competent to make decisions for themselves and those for whom they are designated as having primary custodial responsibility. Consequently, only they can give consent for treatment of the children under their care (Macbeth 2002). Children, by legal definition, are perceived as lacking the necessary competence to give consent, but they have the psychological capacities to voice assent or dissent (United Nations Convention on the Rights of the Child 1989).

In clinical scenarios where safety is the prime consideration, the therapist may need to override the child's right to autonomy. In other situations, where safety of the child is not the issue, the therapist must consider the age, degree of emotional development and cognitive maturation of the child while weighing the degree of respect to be paid to the young client's autonomous decision-making capacity (Sondheimer and Rey 2012).

For example, a depressed adolescent who has not been going to school for a few months, sitting at home the entire day watching TV and playing video games, and who does not see the need or the point of therapy, would need to be motivated to engage in therapy. In contrast, a 17-year-old person with severe specific learning disability who has been brought for therapy, as his parents need the therapist to 'motivate' him to study while he has made an informed decision to pursue vocational training, may not require professional services.

Usually, when parents want treatment for a child and the child assents, treatment proceeds smoothly but often the child and the parent do not see eye to eye on the issue of need for psychiatric evaluation or therapy. Most parents make decisions that they believe are in the best interest of their child. In the Indian setting, some parents believe that the child need not have a say in decisions regarding their own treatment as they are not mature enough to weigh the pros and cons and make an appropriate choice. This is the case especially when the child expresses dissent about a particular treatment that is recommended by the treating team and consented by the parent.

To facilitate collaborative relationships with minors and their guardians, and to enhance patient autonomy, it is the duty of the culturally aware mental health professional to engage in a discussion with the parents. They may be helped to understand that even though the legal need for child's consent is absent, it is important to include the child in making decisions about his/her treatment. This is especially important in psychotherapy as it conveys respect. The therapeutic alliance between the child and the therapist is strengthened when the child feels that his opinion has been given consideration rather than experiencing treatment/therapy as being imposed by others.

3.1.3 Maintaining Confidentiality

Child mental health issues can be understood in the context of interactions between biological predisposition and environmental influences which include family, community and culture. Child psychotherapy cannot be carried out in vacuum without interaction with parents or guardians. Also, parents have the right to be informed about therapy and are entitled to regular updates about the child's progress. Therein lies the challenge of protecting the child's privacy, maintaining confidentiality and keeping the parents involved in the process of therapy.

Consider the following vignette:

Case 3.1

Miss V is a 13-year-old girl with Anorexia Nervosa who has given her assent for cognitive behavior therapy. During one of the therapy sessions, she revealed that for the past 10 days, after every meal at home, she would go the bathroom and vomit. At school, she had not been eating at all. Her parents had no knowledge about the purging behaviour. She tells the therapist that this information must be kept confidential and must not be divulged to her parents at all.

Such a situation presents a veritable dilemma to the therapist as this information needs to be told to the parents, being a potentially life-threatening behaviour which needs monitoring. On the other hand, if it is told then it may result in the adolescent not trusting the therapist any longer. This challenge can be dealt with by ensuring that the following details are shared with the adolescent and the family when a recommendation for therapy is made.

1. Explanation about the rationale and structure, duration and frequency of therapy.
2. Confidentiality in the therapeutic space as well as the limits of confidentiality in a manner appropriate to the developmental level of the child. It is important to describe to the child, the scenarios in which one may need to override confidentiality and disclose information to the parents such as personal neglect, imminent

harm to self or others and high-risk behaviour. Examples of scenarios that mandate disclosure are instances of harm to the child/adolescent—abuse/neglect, suicidal ideation, plan or attempt. Instances of harm to others—homicidal intent or intent to hurt others—physically or emotionally such as bullying and sexual assault. Instances of high-risk behaviour include; planning to run away from home, harmful use of substances, engaging in sexual activity with multiple partners, engaging in potentially harmful online interactions.
   This helps the young person to make an independent decision about disclosing information to the therapist.
3. The child/adolescent should be informed about the frequency of sessions with parents and can be invited to discuss what he/she would like to be communicated in the sessions. The therapist must also offer to review with the child, the summary of the session with his/her parents after it occurs with due respect to the confidentiality of the parents.

In the above scenario, the issue of confidentiality and its limits could have been introduced in the beginning itself in the following manner:

> Before we begin, I would like to discuss with you the "rules" of our meeting. These rules may not be new to you since you have seen therapists before but it's important that we go over them. This is a private space where you can express your thoughts, feelings that you may find difficult to say in front of others. Although this is private and what you talk about will stay in this room, there are exceptions to this. First, if you were to tell me that you were in a danger of some kind, then I would have to tell your parents about that. Secondly, if you were to tell me that another person is in danger of some sort, then I would have to talk to some adults about it. I will do this only after telling you. We will be able to talk about who needs to know and how we should tell them. Does that make sense? Do you have any questions about what I just said?

When the child reveals regarding the purging behaviour in a subsequent session, the therapist discusses with her the serious consequences of low weight and vomiting and how this puts her in danger. A reference to the earlier discussion on confidentiality and its limits needs to be made. Considering the imminent harm to adolescent's health, the need to inform the parents about this to ensure her safety is emphasized. Then the therapist and Miss V could discuss how and when they would tell her parents.

In our country, children live with parents until they get married or even after that in a joint family system. Many parents believe that they need to know everything that is going on in their child's life and may feel offended when they do not know or feel upset that their child trusts the mental health professional more than them. They may feel responsible for the child's problem and feel that they have a 'right' to know and expect the clinician to discuss confidential details discussed in therapy.

It is important for the mental health professional to discuss issues related to child's need for privacy with the parents. They need to be helped to understand the difference between privacy and secrecy. The information shared in the therapeutic space is secret but its therapeutic purpose is to promote freedom in the space and not to hide for any covert purpose (Ascherman and Rubin 2008). They also need to be reassured that they will be met on a regular basis to keep them updated about the progress and that information suggesting imminent danger to the child or others will be told to them at once.

3.2 Therapist-Related Ethical Considerations

3.2.1 Competence

In India, child mental health is a relatively new field and there is no uniform standard to obtain the requisite qualification to practice child psychotherapy. It is widely perceived that there is no definitive harm or serious threat to life in the practice of mental health. Therefore minimal importance seems to be attached to the level of competence needed to be attained by a practitioner. Clinical competency is especially important when working with children, given the greater potential impact on their developing brains and selves and their greater vulnerability (Belitz and Bailey 2009).

The scope of preventative interventions that are possible is tremendous, so it is vital that the services provided be of high standard. In order to meet the needs of these vulnerable children, therapists need a set of distinctive skills and competences.

Active efforts need to be made to have good psychotherapy training programs in India focusing on child mental health with adequate supervision of the trainees enrolled in the program. Areas that might need emphasis are emerging issues such as sexuality in adolescents, trauma, behavioural addiction and therapy for children and adolescents with developmental disorders. Focus is also required in dealing with issues that are entwined with personal/societal values or contentious within the field of mental health such as sexual orientation and gender identity. These programs must also include training in group interventions and interventions in the community especially directed at children in difficult circumstances and children in conflict with the law.

A fundamental principle of medical ethics is that, outside extenuating circumstances, physicians provide only care that they are competent to deliver (Belitz and Bailey 2009). This is applicable to all practitioners in the field of child mental health care as well. In countries like ours where such services are scarce, a general psychiatric practitioner may have to address the needs of children as well. How can mental health practitioners gain competence in these domains or negotiate the complex ethical dilemmas that arise?

A practitioner is obligated to keep in pace with the rapid advancements in theory and practice by regularly engaging in learning opportunities like workshops, distance learning and careful reading of the literature.

In difficult cases, one must engage in constantly revising the therapy formulations as the sessions' progress, and engage in reflection regarding the approaches that have and have not helped in therapy. In cases where one finds that therapy is not progressing, one must be mindful and monitor oneself carefully and take help in the form of consultation/supervision to ascertain an appropriate course of action.

3.2.2 Counter Transference: What Does One Do When One Experiences Strong Feelings Towards a Client?

Countertransference refers to the feelings evoked in the therapist by the patient's transference projections. These can be a useful guide to the patient's expectations of relationships. They are easier to identify if they are not congruent with the therapist's personality and expectation of his or her role (Hughes and Kerr 2000). While working with adolescents, therapists may have to experience defiance, their challenging a therapist's competence, unwillingness to speak and emotional outbursts or intense idealization. These behaviours evoke various emotions in the therapist. These reactions are unavoidable and help in understanding our patients. One must be vigilant towards feelings or behaviour towards the client or client's parents that markedly deviate from one's usual practice. Common warning signs in the behaviour of the therapist include recurrent lateness to sessions, extensions of sessions, touching of the patient, gifts to the patient and contact with the patient outside of scheduled sessions, especially outside of the office setting (Ascherman and Rubin 2008).

Dealing with countertransference The critical element is having a reasonable level of awareness of one's own thoughts and feelings and a grasp of whether these deviate from good professional behaviour. Good practices include:

1. A questioning attitude towards one's own feelings and motives,
2. Recognition that we all have 'blind spots'.
3. An understanding that staff are affected by patients (Hughes and Kerr 2000).

One may take the help of one's psychotherapy supervisor or a senior clinician to help determine the best course of action in the situation. If one is working with a team in treating clients: using the team to clarify what a difficult patient projects into the treatment relationship can contribute to an understanding of the patient's transference to the team or to individuals (Kerr 1999). One may undergo personal therapy to become more aware of one's own unconscious needs and fears. A case elucidates the need to identify countertransference (see Case 3.2).

Case 3.2: The Need to Identify Countertransference

J, a 14-year-old boy, was brought for in-patient care. This was his second admission. He was brought with complaints of defiance, frequent fights with his stepmother, anger outbursts, lying and stealing. During the first interview with the trainee-therapist, he seemed uninterested in talking to her, asked for the therapist who helped him during his last visit—reiterating that she was the 'best' therapist and knew him well. He responded half-heartedly to her questions, saying that it is all written in the file. He was also heard talking to his father, saying that this therapist will not be able to help him at all and that she did not seem to be on his side. This behaviour continued for the next few days and the trainee began to feel that she was not good enough. She felt hopeless about her ability and began to postpone the sessions with the child.

The trainee discussed this case with her therapy supervisor, understanding that this reaction was quite uncharacteristic of her. The discussion helped her realize that the child had been living in a broken home with a mother who had difficulties in emotional regulation and they had an enmeshed relationship. His mother had told him that his father had abandoned the family. The mother had then died unexpectedly and social services had forced the father to take custody of the child. He was wary of his father as he felt that he did not care for him and his mother.

The behaviour towards the therapist can be better understood by the child's fear to form new relationships—fear of abandonment and splitting—idealization of the old therapist and denigration of the new therapist. The child felt very vulnerable and fragile and was afraid of being rejected yet again and he rejected the new therapist instead.

The therapist's awareness of her own feelings and uncharacteristic behaviour, and the supervisory discussion helped her understand her counter-transference and interpreting this helped her understand the patient and his therapy needs.

3.2.3 Understanding Multiple Relationships

The American Psychological Association (2002) states that, "A multiple relationship occurs when a psychologist is in a professional role with a person and (1) at the same time is in another role with the same person, (2) at the same time is in a relationship with a person closely associated with or related to the person with whom the psychologist has the professional relationship, or (3) promises to enter into another relationship in the future with the person or a person closely associated with or related to the person". It means providing therapy in addition to having a personal, fraternal, business, financial or sexual and romantic relationships. The therapist refrains from entering this kind of a multiple relationship as there is a possibility of impairment of one's objectivity, competence and effectiveness. In the case of children and adolescents, this could manifest in a close relationship with the child or with the parents of the child.

In case such an overlap is unavoidable, then it is the therapist's responsibility to define the therapeutic relationship and the principles of confidentiality and explain to the family that the social relationship will be affected and will need to be limited henceforth.

3.2.4 More Than One Therapist

Children and adolescents find it difficult to be engaged in concurrent psychotherapy sessions with more than one therapist (Ascherman and Rubin 2008). This also increases the chances of the child receiving confusing messages from the therapists. There is another possibility that the therapist who is discussing the core issue may be devalued and the therapist who is farther way from the issue is idealized and both therapists do not get a holistic picture of the patient and his problems.

There is a need for two therapists in cases like eating disorders where one person needs to take care of the problem behaviour and another person takes an exploratory approach and works on overall growth and development. In some settings, one therapist works with the child and the other works with the parents. Both clinicians should maintain awareness that they are vulnerable to distorted views of the child and parents influenced by what they hear or experience in their respective processes (Roberts et al. 2002).

There must be ongoing communication between the therapists without breach of confidentiality. This will help both of them gain an understanding of the overall picture and to ensure coordination in therapy.

3.3 Online Practices

Child and adolescent clients are often 'digital natives' and it may be a natural tendency for them to conduct online searches about their therapists and attempt to 'friend' them on social networking sites.

Online self-disclosure by the therapist may cause the client's perception of the relationship to be a more casual or even social one. This may violate the boundaries or context of therapy as a sanctuary for exploring personal issues (Kaslow et al. 2011). It is particularly important to set appropriate boundaries with clients. To do this, a psychologist may need to create and maintain a formal social networking site policy as part of the informed-consent process Such policies could lay out psychologists' expectations for using such sites, namely that practitioners do not 'friend' or interact with clients on social networking sites (Kolmes 2010). Practitioners should also inform clients that they do not search for them online, unless the client has given consent.

Therapists must be conscious about the information clients can see online. Therapists must search themselves online periodically so that they are aware about the details that the clients may find. Therapists should proactively set controls that limit who see their personal information. Practitioners could set security levels on social networking sites as high as possible, allowing for access for friends only (American Medical Association 2010).

Therapists must assume that everything they post online may be read by their clients. Therapists must be careful when they post case studies online and must make efforts to change identifying information significantly and protect their client's right to privacy.

3.4 Documentation

Documentation seems like a laborious and meaningless chore but it plays an important role in helping us reflect and review our therapeutic interventions. It helps us weigh alternative treatment approaches and monitor efficacy of treatment. It serves as a record of our work; it can help in maintaining continuity of care and facilitates coordination of efforts by enabling communication among team members. From the legal viewpoint, documentation is the best way of protecting oneself in the face of ethical charges.

The details that must be recorded as per the APA resource document (American Psychological Association 2002) on record keeping of therapy are:

1. Identifying information and first contact.
2. Relevant history and risk factors, medical status and attempts to get prior treatment records.
3. Dates of service and fees.
4. Diagnostic impressions, assessments, treatment plans, consultation summary and testing reports and supporting data, and progress notes. Include not only the treatments chosen, but treatments considered and rejected.
5. Informed-consent documentation, consent to audiotape or videotape, and release of information documentation.
6. Relevant telephone calls and out-of-office contacts.
7. Follow-up efforts when clients drop out of therapy.

Psychotherapy notes: The therapist may use his or her judgement about what to write in an official medical/psychiatric record in order not to expose the patient to a breach of privacy and confidentiality that would undermine the psychotherapy and harm the patient. This is done considering the vulnerability of medical records to unauthorized access. Concise documentation of psychotherapy could be done in the general psychiatric record. It could include details of psychotherapy like the date, duration of the session, psychotherapeutic intervention and a brief mention of major themes or topic(s) addressed.

The therapist may maintain a separate detailed account of the content of therapy, process interactions and sensitive information about the patient and other individuals in the patient's life. The practitioner's personal reactions, hypotheses or speculations may also be recorded in the psychotherapy file. All efforts must be taken to ensure adequate confidentiality of this document (see Case 3.3).

Case 3.3: An ethical dilemma in documentation

A 14-year-old girl was referred to the child psychiatry outpatient services for evaluation following a history of child sexual abuse (CSA). She was eight months pregnant. During therapy, it was revealed that the perpetrator had been in an extramarital relationship with the mother. The mother had not known about the ongoing abuse of her child. When she became aware about the pregnancy of her child and that the doctors opined that an abortion was not possible, she made a suicide 'pact' with the girl. Mother and daughter consumed poison but were unharmed. This happened two months prior to the consultation, seemed like an impulsive attempt and the mother and the child did not have any symptoms suggestive of depression.

The ethical dilemma: The hospital file could be asked for perusal in the court of law. Considering the fact that the mother is the only caregiver for the child now and that child must get justice, should the mother's relationship with the perpetrator be documented in the file? Should the fact that the mother made a suicide pact with child be documented? Should legal action be taken on the only caregiver that the child has, thus leaving her in the care of child protective services?

In such a situation, it is important to consider the vulnerability of the record to (mis) interpretation. One way to circumvent this problem is to note the details of the sexual abuse, its impact on the child and a brief summary of sessions in the psychiatric record. The other details which were elicited during sessions with the child could be a part of the child's psychotherapy notes, maintained separately as confidential records. In this way, additional details of the family, not in the file, need not be presented at court proceedings thereby preventing the risk of legal action against the child's only caregiver.

This approach is not unethical because it is based on two premises: (i) the child's interests which are served by protecting and maintaining the only caregiver she has; (ii) the key issue to be addressed is the child's sexual abuse, not the mother's marital issues or relationship engagements, i.e. the nature of the perpetrator's relationship to the mother bears no relevance to prosecuting him for abusing the child. Also, in an ideal scenario, the records or psychotherapy notes of the child should remain separate from notes of family sessions or notes of meetings with one or both parents. Keeping separate files or file sections for each party affords confidentiality for all the parties involved (Koocher 2008).

3.5 Communication with Third Parties

Requests for information about a child/adolescent seen in psychotherapy are numerous. The therapist may get requests from agencies that refer children for services like Child Welfare Committee, government state homes and orphanages. Schools ask for details of children's problems, progress in therapy and help in dealing with them at school. The police or the court may ask for information in child custody cases and cases related to juvenile offenders and child sexual abuse.

Clients enter the psychotherapeutic relationship with confidence that one's privacy will be protected. The privilege of permitting release of this confidential information, in the case of children and adolescents, lies with their parents. It is also the duty of the mental health professional to protect the child client's confidentiality.

When the therapist receives a request for release of information, it must be reviewed carefully. One must consider the appropriateness of the request, the necessity for the information and the impact of release on the child, family and the therapy process. One must discuss the request with the child and the family, and then the decision is made regarding communication with the third party. The therapist must assess the goals of the request and prepare a report. The report must be prepared with care and the minimum necessary information that is relevant to the goals of the request must be given.

3.6 Child Therapy, Ethics and Law

Two important areas where therapy, ethics and law are entwined are in child sexual abuse reporting and child custody disputes.

3.6.1 Child Abuse and Reporting

Section 19 of Protection of Children from Sexual Offences (POCSO) Act, 2012 states that (1) "Notwithstanding anything contained in the Code of Criminal Procedure, 1973, any person (including child) who has apprehension that an offence under this Act is likely to be committed or has knowledge that such an offence has been committed, he shall provide such information to the Special Juvenile Police Unit; or the local police" (Ministry of Women and Child Development 2012).

This means that any person including mental health professionals are required to bring the matter to the notice of the police as soon as the incident comes to his/her notice even if the survivor does not give consent to reporting the case.

Case 3.4: Child Sexual Abuse Reporting
A 16-year-old girl presented to the child psychiatry outpatient department with history of lying, stealing, defiance at home and poor scholastic performance. Family history revealed that the parents were separated from each other for the past two years with the father having applied for divorce and the mother for restitution of marriage. This case was ongoing and the mother and daughter were staying together and the father was living separately. A provisional diagnosis of conduct disorder in the home context was made and the child was taken up for outpatient-based therapy. During the course of the therapy, the child revealed that the father had sexually abused her (contact non-penetrative in nature) three years ago. This had occurred once. She did not want this to be revealed to the mother or be reported to the police. She hoped that her parents would get back together. She strongly believed that her father would not do it again and that she knew how to protect herself if it happened again. She also knew that reporting this may cause her father to go to jail and shatter her hopes of the family reuniting.

Facts to be considered

(1) The abuse occurred three years ago.
(2) If child abuse is reported to the mother or the police, it will be a breach of confidentiality, the child's trust will be broken and she may not come for therapy.
(3) The action taken by the Child Welfare Committee and the police may result in legal action against the father.
(4) In case the child abuse is not reported and the family court orders the reconstitution of marriage of the parents, the child will be at risk of abuse.
(5) Reporting laws in India.

A therapist must inform the child at the beginning of therapy the limits to confidentiality so that the child/adolescent makes a choice about what they wish to reveal in therapy. In the case mentioned in Case 3.4, the therapist will need to remind the child the need for mandatory reporting on the part of the therapist. It is also the therapist's responsibility to help the child understand the various consequences that may take place after reporting. The therapist must refrain from giving false reassurances and answer the child's questions to the best of his/her ability. If the therapist does not know the answers to some of the questions, this must be clearly stated.

Weinstock and Weinstock (1988) assert that child abuse laws have a strong impact upon clinicians and present a serious threat to therapeutic confidentiality. The positive aspect of mandatory reporting is that it helps to bring more cases of child sexual abuse to attention. However, the negative effects may include: the detrimental effects due to fears of liability upon therapy, over reporting, a deterioration of patient-therapist trust and disregard for the effects of reporting upon victims and families.

The downside is that mandatory laws, on rare occasions, may prohibit the professional from using one's discretionary judgement to act in the 'best interests of the child', especially in cases where reporting may cause more harm than good to the child and the family.

3.6.2 *Child Custody Disputes*

It is advisable to seek consent of both parents when agreeing to see children whose parents have separated or divorced (see example in Case 3.5). If the parent bringing the child for treatment hesitates about allowing the other parent to know of and consent to treatment, the therapist should not agree to take on the case. If told that the other parent's whereabouts are unknown, that their parental rights have been terminated or that they are incarcerated, then therapists should document those assertions in the case file (Koocher 2008).

Case 3.5: Child Custody Disputes
A 7-year-old boy was referred for play therapy for emotional disorder. The mother reported that the parents are separated and filing for divorce.

It is possible that the therapist may get embroiled in a dispute over custody of this child in the future.

A few precautionary steps may minimize the risks of ethical violations. In case of impending divorce, the therapist may ask both the parents to give in writing that the child's therapist will not be called as a witness and that they will not seek child's psychotherapy records as evidence. This is done as providing evidence in litigation involving child clients has a potential to disrupt the therapeutic alliance and compromise the treatment process. If the therapist is called to testify in court, then one must limit one's comments to treatment of the child. One must refrain from making any recommendations regarding custody of the child as it may be beyond the scope of one's work.

In case the parent feels that the child's psychological condition needs to be considered in the custody dispute, they can be directed to seek an evaluation from an independent evaluator. This facilitates an objective evaluation with attention to all the concerned factors.

3.7 Termination

According to the American Psychological Association (2002) code of ethics, therapists can end therapy when the therapy goals are met, clients are not benefiting or getting harmed by therapy and when there is a threat to the therapist, client or others (APA 2002).

It is advisable that a treatment timeline is given to the family in the first few sessions itself. It is the therapist's responsibility to recognize the signs that therapy has attained a reasonable level of success. When the goals set at the beginning of therapy are achieved, then termination could take place. The clients must be involved in the plan of termination. In pre-termination counselling, one needs to review gains made in treatment. Potential relapse risks and ways of handling them need to be discussed. Therapists should offer continued interest and availability, in cases of emergency (Polemikos and Papaeliou 2006).

It is vital for one to know the distinction between termination and abandonment. Abandonment is considered an inappropriate termination when a client's ongoing treatment needs are not adequately addressed by the psychotherapist, either when treatment ends or during the course of treatment due to unavailability (Vasquez et al. 2008) (see example in Case 3.6).

Case 3.6: Termination of Therapy
Miss Z is a psychiatry trainee who had a difficult, challenging family for therapy for three months. She did not seem to be making any progress. She was glad that the child psychiatry posting was getting over and she did not have to see the patient anymore. She hoped that the case would 'automatically' be given to some other therapist and the patient's care would continue.

Patients and their families experience a sense of abandonment when unplanned events occur and they feel unprepared to face the possibilities. In this case, the trainee should have informed the family about the end of her posting in advance. She could reassure the family that another trainee would take her place and that appropriate care would continue. It is the supervisor's responsibility to ensure that arrangements are made for continuity of care.

Even when termination is sudden, like the family moving out of town, a session must be devoted to helping the child understand reason for termination and work on feelings related to it.

When a client does not turn up for appointments, it is tempting to feel that one's obligations have ended as well. Therapists must make attempts to contact clients by telephone. In case that fails too, it is recommended that a letter be written to them. This could have details of the assessment, treatment needs and recommendations.

3.8 Showing Feelings in Therapy: Crying?

There is dearth of the literature on therapists crying in therapy and more so in therapy with children and adolescents. In a recent study about therapists crying in therapy, 72 % of respondents reported having cried with their patients. Despite this, there is no clear consensus among mental health professionals about the appropriateness of crying. The study found that years of experience and therapy style were the two major predictors of whether a therapist had ever been moved to tears as part of a therapy session. Most therapists who reported crying did not feel like their tears had adversely impacted the working alliance, and, in fact, the majority felt like it had strengthened the relationship they had with their patient (Blume-Marcovici et al. 2013). Tears may be appropriate and even potentially helpful in the background of a strong therapeutic alliance.

3.9 Public Encounters

Public encounters with the patient pose a challenge as acknowledging them in public may be a violation of their privacy and confidentiality. In general, it is best for the therapist to explain to one's patient and the parents that, in the case of an encounter outside of the office, one will err in the direction of not acknowledging them unless they initiate an acknowledgment (Ascherman and Rubin 2008). When this does happen then it is best for the therapist to respond in a socially appropriate manner without encouraging further discussion. In the subsequent session one must explore the patient's reaction to the encounter.

3.10 Gifts

Giving and accepting gifts are an expression of affection in relationships. A therapeutic relationship is different from a normal relationship because it is meant to satisfy the psychological needs of only the client. Therefore therapists need to identify the motives of children and identify why gifts are being given.

When a child offers a gift spontaneously, then the therapist must accept it in a graceful manner as they may feel hurt if refused. These gifts may be displayed and this makes children feel that they are valued, helping in improving self-worth. If the gift is from the parents, then it is best to appreciate the gesture and politely decline.

Therapists may also give gifts to children as reinforcement in behavioural approaches. These should be of low monetary value, preferably consumable and relevant to the program.

3.11 Fees

Fees help clients understand that the relationship is a professional one and helps them understand the importance of the process. The child should be informed that the parents are paying for the service of the therapists. The fee schedule must be discussed in the first session itself and conscientious collection of fees must be done. In case there is failure to pay fees, a meeting with the parents must be set up at once to address the issues.

3.12 Conclusion

Therapy with children and adolescents is challenging, exhausting and immensely rewarding. Working with this population presents complex ethical dilemmas and confusions. There are international ethical guidelines prescribed by American Academy of Child And Adolescent Psychiatry (AACAP 2009) and by International Association for Child and Adolescent Psychiatry and Allied Professions (IACAPAP 2006). In the Indian context, general ethical guidelines for psychiatrists (Agarwal and Gupta 1999) and psychologists (National Academy of Psychologists 2010) exist but they do not pertain to treatment of children in particular. De Sousa (2010) has comprehensively reviewed ethical issues pertaining to management of children with mental health problems in our country. These guidelines possess key components that can help us navigate these complex challenges that therapeutic work with children brings.

Despite these guidelines, therapists must understand that these one cannot act in stereotyped ways in therapy. Psychotherapists must embrace this uncertainty with the anxiety that this brings. They should be able to deal with this uncertainty of professional dilemmas by being aware of the ethical principles. They must use this ethical awareness to assess the situation, take into account all the relevant factors and use their discretion and judgement to come to a decision. One must also be willing to take responsibility for consequences of the decision in a professional manner.

Mental health professionals need to be aware that children represent one of the most vulnerable populations. They must possess an unwavering commitment to their safety, dedication to act in their best interests and be mindful of their rights. This conveys a basic respect for the patient and is the cornerstone of therapy with children and adolescents.

References

Agarwal, A., & Gupta, S. C. (1999). Ethics in Psychiatry. In J. Vyas & N. Ahuja (Eds.), *Text book of post graduate psychiatry* (pp. 1019–1025). New Delhi: Jaypee.

American Academy of Child and Adolescent Psychiatry. (2009). *Code of Ethics*. Retrieved from https://www.aacap.org/App_Themes/AACAP/docs/about_us/transparency_portal/aacap_code_of_ethics_2012.pdf

American Medical Association. (2010). *American Medical Association Policy: Professionalism in the Use of Social Media*. Retrieved from http://www.ama-assn.org/ama/pub/physician-resources/medical-ethics/code-medical-ethics/opinion9124.page?

American Psychiatric Association. (2001). *The principles of medical ethics: With annotations especially applicable to psychiatry*. Washington DC: APA.

American Psychiatric Association. (2002). Documentation of Psychotherapy by Psychiatrists. Retrieved from http://medicareservices.us/psychiatricmednessitynotes.pdf

American Psychological Association. (2002). Ethical principles of psychologists and code of conduct. *American Psychologist, 57*(12), 1060–1073.

Ascherman, L. I., & Rubin, S. (2008). Current ethical issues in child and adolescent psychotherapy. *Child and Adolescent Psychiatric Clinics of North America, 17*(1), 21–35.

Belitz, J., & Bailey, R. A. (2009). Clinical ethics for the treatment of children and adolescents: a guide for general psychiatrists. *Psychiatric Clinics of North America, 32*(2), 243–257.

Blume-Marcovici, A. C., Stolberg, R. A., & Khademi, M. (2013). Do therapists cry in therapy? The role of experience and other factors in therapists' tears. *Psychotherapy, 50*(2), 224.

de Sousa, A. (2010). Ethical issues in child and adolescent psychotherapy: A clinical review. *Indian Journal of Medical Ethics, 7*(3), 157–161.

Gaines, R. (1995). The Treatment of children. In M. Lionells, J. Fiscalini, C. H. Mann, & D. B. Sterm (Eds.), *Handbook of interpersonal psychoanalysis* (pp. 751–769). Hillsdale, New Jersey: Analytic Press.

Gaines, R. (2003). Therapist self disclosure with children, adolescents, and their parents. *Journal of Clinical Psychology, 59*(5), 569–580.

Hughes, P., & Kerr, I. (2000). Transference and countertransference in communication between doctor and patient. *Advances in Psychiatric Treatment, 6*(1), 57–64.

International Association for Child and Adolescent Psychiatry and Allied Professions. (2006). *IACACAP Ethics 2006*. Retrieved from http://iacapap.org/wp-content/uploads/IACAPAP-Ethics-2006.pdf

Kaslow, F. W., Patterson, T., & Gottlieb, M. (2011). Ethical dilemmas in psychologists accessing Internet data: Is it justified? *Professional Psychology: Research and Practice, 42*(2), 105.

Kerr, I. B. (1999). Cognitive analytic therapy for borderline personality disorder in the context of a community mental health team: individual and organizational psychodynamic implications. *British Journal of Psychotherapy, 15*(4), 425–438.

Kolmes, K. (2010). My private practice social media policy. Retrieved from http://www.apait.org/apait/resources/articles/kolmessocmed.pdf

Koocher, G. P. (2008). Ethical challenges in mental health services to children and families. *Journal of Clinical Psychology, 64*(5), 601–12. http://doi.org/10.1002/jclp.20476

Macbeth, J. (2002). Legal issues in the treatment of minors. In D. Schetky & E. Benedek (Eds.), *Principles and practice of child and adolescent forensic psychiatry* (pp. 309–323). Washington DC: American Psychiatric Publishing.

Ministry of Women and Child Development. (2012). *The protection of children against sexual offences bill*. Retrieved from http://wcd.nic.in/sites/default/files/childprotection31072012.pdf

National Academy of Psychology India. (2010). *Ethical principles for psychologists*. Retrieved from http://naopindia.org/ethical-principles

Polemikos, N., & Papaeliou, C. (2006). Some practical issues in working integratively with children and adolescents. In E. O'Leary & M. Murphy (Eds.), *New approaches to integration in psychotherapy* (pp. 116–128). London: Routledge.

Roberts, L. W., Geppert, C. M. A., & Bailey, R. (2002). Ethics in psychiatric practice: Essential ethics skills, informed consent, the therapeutic relationship, and confidentiality. *Journal of Psychiatric Practice, 8*(5), 290–305.

Sondheimer, A., & Rey, J. (2012). Ethics and international child and adolescent psychiatry. In J. Rey (Ed.), *IACAPAP e-textbook of child and adolescent mental health* (pp. 1–27). Geneva: International Association for Child and Adolescent Psychiatry and Allied Professions.

Sullivan, H. S. (1940). Conceptions of modern psychiatry: The first William Alanson White Memorial Lectures. *Psychiatry, 3*(1), 1–117.

United Nations Convention on the Rights of the Child. (1989). Retrieved from http://www.ohchr.org/en/professionalinterest/pages/crc.aspx

Vasquez, M. J. T., Bingham, R. P., & Barnett, J. E. (2008). Psychotherapy termination: Clinical and ethical responsibilities. *Journal of Clinical Psychology, 64*(5), 653–665.

Weinstock, R., & Weinstock, D. (1988). Child abuse reporting trends: an unprecedented threat to confidentiality. *Journal of Forensic Sciences, 33*(2), 418–431.

Chapter 4
Ethical Issues in Working with Children in the Consultation-Liaison Context

Vijaya Raman

Abstract Children needing input from a mental health professional in a general hospital setting is on the increase in India, with the advances in awareness among paediatricians and the willingness of parents to follow through on a referral that is made. In this context, unique ethical issues while working with this group are emerging and need to be kept in mind while planning intervention (APA 1990). Some general ethical principles when working with children are relevant to the consultation-liaison setting as well—issues regarding confidentiality, informed consent, respecting boundaries, data recording and protection of data, communication with the stake holders and considerations when using email or the telephone. In addition, some unique issues are relevant to the general hospital setting. This includes issues with documentation—access of medical records to multiple medical professionals, the physical setting where the therapeutic work is undertaken—sometimes at the bedside, sometimes in the corridor, and other times, at the intensive care unit. Issues with regard to time constraints—there may just be one session before discharge, consultation with the physician in charge—who may or may not be sensitive to the psychological needs, are all relevant. In addition, child variables—physically not being able to handle long therapy sessions, not having a say in the decision-making and family variables—multiple decision-makers and lack of psychological sophistication all need to be kept in mind. All the above issues are discussed in detail and an overview of its relevance and utility in the Indian setting is provided.

Keywords Ethics · Therapy · Children · Consultation-liaison context

V. Raman (✉)
Department of Psychiatry, St. John's Medical College Hospital, Bengaluru, India
e-mail: vraman12@yahoo.co.uk

P. Bhola and A. Raguram (eds.), *Ethical Issues in Counselling and Psychotherapy Practice*, DOI 10.1007/978-981-10-1808-4_4

4.1 Introduction

Engaging in psychotherapy with children and families is challenging in itself due to complexities in relating to the child and working with the family dynamics all the time. In addition, planning psychotherapy in an ethical and legal manner is a bigger challenge. Ethical guidelines are most often written for adults and extrapolated to children. And, in the process, the essence of working with this population may be lost.

Psychotherapy with children requires the knowledge of developmental issues, standard of care practices, applicable laws as well as ethical principles. Child and adolescent mental health professionals encounter ethical situations that often differ from those of general mental health professionals. Several factors increase the complexity of ethical practice in child and adolescent mental health including the variety of persons involved in meeting the mental health needs of children, the care of a vulnerable population whose welfare relies on others, the limited access to mental health services for children, and the advancements in science which provide an increasing number of diagnostic and treatment options (Ascherman and Rubin 2008).

The practical and ethical implications of working with children are substantial. Adults unavoidably look down on children, and children unavoidably look up to adults. Being smaller, they are more readily intimidated by adults. These simple physical facts both establish and illustrate the relative imbalance of power, influence, and authority between children and adults.

In the consultation–liaison context, like in any other context, the treatment of children is based on several principles (Schetky 1995a, b).

1. Do no harm to the child,
2. Do what is in the best interest of the child,
3. Always strive to protect the privacy of the child's communications,
4. Respect the child as well as the family regardless of race, religion, socioeconomic status, education or intellectual level, and
5. Promote and support the highest level of development and autonomy in the child.

Common ethical principles such as 'beneficence' and 'do no harm' often are more multifaceted and less straightforward. As a salient example, children rarely refer themselves for psychiatric care. Typically, they are brought into care by others, either because of genuine concern for their wellbeing or because they are causing trouble for others. The treatment objectives expressed by those who refer the child may be quite different from the child's goals. Indeed, the child may have no wish for treatment, creating a difficult environment in which to develop a motivated treatment contract with the child.

Psychotherapists are ethically compelled to respect and integrate the child's expressed wishes, the referring party's expressed outcomes, the child's maturational needs, and the family's cultural beliefs about family roles and child-rearing practices. General psychiatrists have the relative luxury of focusing solely on the adult patient, whereas a child psychiatrist must consider the needs, abilities, and availability of the child's parents, family, and school. Such considerations obviously bring new ethical considerations and challenges to the treatment.

Psychological work with emotionally disturbed children who also have physical illnesses is very challenging, both clinically and, many a time, ethically as well. Referrals to the clinical psychologist vary in their presentations and the moral/ethical/legal and medical implications all need to be considered in different degrees with every child and family. The issues are more complicated by the obvious focus on the current medical condition.

As a rule, children have many more expectations and mandates placed on them by society, have a more limited set of adaptive abilities, and are granted fewer behavioural options than adults. For example, adults have a relatively broad range of vocational options available, each with their own set of advantages and demands; children, however, must go to school. Adults may avoid certain behavioural or ethical demands simply by avoiding the circumstances in which they arise; children often are unable to avail themselves of such environmental and behavioural flexibility.

4.2 General Ethical Considerations in Psychological Intervention with Children

The practice of psychotherapy, whether at the bedside or in the office room or in the clinic, has several common factors and several unique challenges. Establishing rapport with the child and with the parent/guardian is crucial for any intervention to work. It is also important to make the child feel that the entire process of therapy is with the child as the primary client.

Since children develop within a unique system that consists of many individuals —siblings, parents, teachers, friends, and clinicians and many organizations— school, day care, clinics, etc., the needs of all the stakeholders need to be taken into account.

Some of the ethics-based rules that apply to the practice of child and adolescent psychiatry are clear and generally agreed upon (Schetky 1995a, b). For example, rules against sexual contact or harsh and abusive treatment are encoded as boundary violations. They are based on the recognition that such experiences cause trauma, further distorting and injuring the child's trust, self-esteem, and capacity for intimate relationships.

The core ethical issues when caring for children are (Belitz and Bailey 2009):

1. Children are inherently more vulnerable than adults.
2. Children's abilities are more variable and change over time.
3. Children are more reliant upon others and upon their environment.

4. Ethical principles and practices in the treatment of adults must be modified in response to the child's current developmental abilities and legal status.
5. Boundary and role issues are often more complex when caring for children than for adults.
6. Adult psychiatric practices, and the adult psychiatric knowledge base, do not transfer reliably to the care of children.
7. Practitioners must develop skills to work with families, agencies, and systems.
8. It is important to monitor one's own actions and motivations.
9. Seeking consultation and advice is helpful in difficult situations.
10. It is essential to maintain an absolute commitment to the safety and wellbeing of the child.

The common ethical issues in child and adolescent mental health care are the following:

4.2.1 Confidentiality

As applied to child and adolescent psychiatry, the principle of confidentiality focuses on the patient's right to have information kept private and confidential. The American Association of Child and Adolescent Psychiatry (AACAP 2009) Code of Ethics states, "Certainty that their verbal expressions are protected as confidential allows minor patients to reveal their feelings and thoughts to the clinicians providing care, with the assurance that the contents of their discussions will not be communicated to others without their permission".

Fidelity and autonomy are linked ethical concepts that are relevant to confidentiality when applied in child and adolescent psychiatry, in that 'fidelity' refers to upholding one's duty and loyalty to a patient, while 'autonomy' refers to the child's (and especially the adolescent's) right to self-determination within a developmental context.

4.2.2 Informed Consent

Upon recommending a psychotherapy process, the child and adolescent mental health professional has the responsibility and opportunity to review with parents or guardians the structure of the psychotherapy frame, and most importantly, its rationale. In addition to discussing schedule, duration of sessions, frequency and estimated duration of treatment, issues relevant to the child's privacy should be addressed. Consent needs to be taken from both the child and the adolescent.

4.2.3 Respecting Boundaries

During any psychotherapy process, subtle instances can arise that pose risk for boundary issues. With younger children, the issues may include wanting to sit on the lap or give a hug. Other children may want to invite the therapist to their birthday parties or other functions at home. Adolescents may challenge boundaries by posing personal questions, such as the therapist's marital status or how many children she/he has etc. Such questions indicate curiosity and are potential opportunities to understand the child further. Motivations for the questions can be explored without necessarily having to provide answers.

4.2.4 Data Recording and Protection

It is the responsibility of the mental health professional to protect information about the child and his/her family. The duty to hold in confidence that which is revealed in the context of a doctor–patient relationship is a very responsible one. All notes and any other manner of recording data should be protected and only released with permission from the child and family.

4.2.5 Communication with Parents and Guardians

Those conducting psychotherapy with children and adolescents encounter unique challenges in their efforts to protect their patient's privacy. It is not possible to conduct therapy in a vacuum sealed from interaction with parents and guardians. When this does occur, it can reflect a parent's trust in the process and respect for their child's privacy or, more ominously, a lack of interest in his or her child's emotional life and an implicit delegation of responsibility for the child's wellbeing on the mental health professional.

4.2.6 Written and Verbal Communication with Parties Outside of Family

Requests for information about a child or adolescent in psychotherapy can be abundant. The therapist becomes the gatekeeper of information and guardian of his or her privacy. It is necessary that the therapist responds to any request for release of information with consideration of its appropriateness, necessity and the potential impact of its release on the child, the family and the therapy process. It is especially necessary to be circumspect when giving a written report as these often contain abundant information beyond the needs that generated the request.

4.2.7 *Ethical and Clinical Considerations with Email and Telephone Communication*

In the modern technologically advancing era, communication requests via email/telephone/Facebook/other messaging systems need to be anticipated and dealt with. There are three main ethical concerns of Internet and email—problems inherent in the mechanics of email-privacy and confidentiality issues and the loss of essential elements of the therapeutic action associated with the therapist–client relationship. Email communication in exchange for direct communication in the office deprives the therapist of critical communication related to facial expression, body language, and voice tone (Kassaw and Gabbard 2002). For similar reasons, the use of the telephone as a regular alternative to direct face to face sessions is also discouraged.

4.2.8 *Unique Issues in the Consultation-Liaison Setting*

When working with children and families in the medical setting, there are a few challenges faced by the mental health professional that are over and above the issues in dealing with them in a general psychiatry setting (Krener 1995; Wright and Roberts 2009). The salient ones relevant to the general hospital setting are enumerated below.

4.2.8.1 Documentation

In a general hospital setting, children usually need to visit various departments for their various problems. As a psychologist working in such a setting, the challenge is to decide what, how, when, and where to write one's notes. Obviously, details about confidential issues need to be recorded separately where access is limited to the stake holders only.

4.2.8.2 Physical Setting

Very often, a mental health professional working with the child team is asked to screen, evaluate, and manage a child in the ward and at the bedside. There are definite issues with confidentiality and the extent of rapport that is possible in such a situation. In India, where extended families usually provide support during inpatient stay, excluding them from discussions can be resented, at best and treated with hostility, at worst. Arranging for a private corner in the ward itself and requesting the staff nurse to ensure that the discussion is not disturbed can be planned easily.

4.2.8.3 Time Constraints

As many mental health professionals working in a general hospital would agree, time constraints play a major role in the extent of involvement possible in the treatment. Many times, referrals are made on the day or the day prior to discharge and keeping the family in the hospital after the medical team has offered discharge can be a challenge. Often, therapy is initiated in the ward and further appointments need to be scheduled for post discharge visits. Working with the treating team to ensure that referrals are made giving some time for evaluation and making the treatment plan by the psychologist would be recommended. Usually it is not the lack of importance to mental health issues that leads to the last minute referrals but more the lack of awareness of the time required and the process of a psychological evaluation.

4.2.8.4 Physician Variables

Given the fact that the treating team has its goals clearly focused on the presenting physical symptom, the shift in emphasis can be a challenge. A lot of times, discussions on clearly confidential information may happen on the corridor, on the bedside, and sometimes over the phone and one needs to bear in mind that a conscious effort may sometimes be needed to get everyone on the treating team to become sensitive to the ethical issues involved in this sharing of information.

4.2.8.5 Therapy Variables

The form, content, and process of therapy is very different in a setting where the primary problem may not be the psychological one. We may need to be flexible in the way the therapy is formatted to suit the needs of the child, family, and the reason for referral. Often, more than one family member may need to be involved in the therapeutic alliance. The frequency of sessions, the duration of sessions, and the therapeutic stance all may need to be tailored to individual needs. Daily sessions, shorter duration of each session and an eclectic approach are all more effective.

4.2.8.6 Child Variables

When a child in a hospital setting is referred for psychological evaluation, the child may be physically not well enough to handle long therapy sessions. Sometimes, being in the hospital itself is so stressful for most children, that they are not really very cooperative for verbal therapies. Rapport building in this context will require a flexible, individually tailored approach that can only be arrived at through experience. The child may have intravenous treatment going on, may have attached monitors, and may also be isolated in the ward and all of these are not conducive to an ideal therapeutic setting. Including the child in the decision-making process is not usually practised in our setting although it is the right thing to do.

4.2.8.7 Family Variables

When we are dealing with children, it is impossible not to deal with the family as well. This is especially the case in India where almost all decisions regarding the child's health, including mental health, are made by the family. This may also include extended family members like grandparents, aunts, uncles, cousins, etc. It is a unique challenge in our setting to draw the line to decide who all need to be involved in the care of the child.

4.2.8.8 End of Life Care

There are numerous challenges and conflicts in terminal situations including some medical as well as psychological issues:

- To take a decision to withdraw advanced life support, nutritional support, or resuscitation.
- To avoid prescribing futile investigations and futile therapies.
- To deal emotionally with the situation of patients without taking sides—of either the treating team or the family.

4.2.8.9 Health Professionals' Conduct

Sometimes when working as part of a multidisciplinary treating team, one may come across several conundrums that have ethical implications:

- Disagree with the treating team about the indication for procedures.
- Witness inappropriate attitudes of colleagues.
- Disagree with breach of confidentiality, inappropriate use of investigations, or medication.

In this context, another important issue that may arise would be when a mental health professional feels compelled to take up an important component of treatment that really comes in the purview of the medical professional's care. In this kind of scenario, the suggestion would be to gently urge the medical professional to deal with the issue while, at the same time, be available to the child and family to fill in the gaps. Sometimes, working in tandem will be an even better option.

4.3 Case Scenarios

Case 4.1
A 4-month-old infant was referred to the clinical psychologist with a view to counsel the parents about the need to take the child off the ventilator as the child had been ventilator-dependent for the past 3 months. The parents were seen by the treating team as being unable to take this decision despite their counselling.

In Case 4.1, the sessions with the parents were planned on a daily basis as they had to come to a decision quickly. Over seven sessions, the parents' concerns became clear. They were unable take a decision to take the child off the ventilator which would mean the child may not survive for too long after that. The onus on them was making them feel distressed. They also shared some concerns about the medical treatment the child had received in the hospital and requested me not to divulge this to the treating team. The parents also reported that they felt the child sometimes responded when spoken to and they felt it was a positive indication and could not turn off the ventilator.

The ethical issues that I encountered here were—'Who is my client?'—Obviously the infant was not able to participate in the decision-making, so the client would have to be the parent/s. The other issue was whether I should share the information about the treatment that the parents were not happy with. When the parents expressly told me not to, I had to respect their wishes. The treating team felt the child was going to have developmental issues and I did share this with the parents. They wanted to know what I would have done in the same situation. Obviously, they had to take an informed decision and that is what I discussed with them. They finally decided to take the baby home on a home ventilation system till their finances lasted.

Case 4.2
A 10-year-old girl was referred by the plastic surgery department from the Burns ward. She had suffered 50 % burns, but, due to some delay in reaching the hospital, infection was setting in. She had been left at home alone for a

few minutes as the mother had to run an errand and the child's clothes had caught fire from a lamp lit in the prayer room. The mother was inconsolable as she felt very guilty at having left her alone at home. At the outset, the doctors had warned the family about the poor prognosis.

For Case 4.2, the sessions were had in the counselling room in the Burns ward with the parents and at the bedside wearing mask and gown with the child. Although I had only four sessions over four days before the child died, I could really communicate with her coherently only on the first day as she was in and out of consciousness on the other days. She did ask me whether she would live several times on the first day. She appeared frightened to be by herself in the ward and swathed in bandages.

Sessions with the mother served to allow her to vent her feelings and share her sorrow. She felt she could not share with her husband as he was being 'practical and unemotional'. She would repeatedly ask the medical team for hope. When the child finally died, she was just not ready to accept it.

The ethical issues in this case were—should I tell the child the truth? Although it was very difficult, after speaking to her parents, I did convey this to the child with her mother by her side. The child appeared to understand what was conveyed to her. She was, however, in no physical state to plan closure. I was left with an uncomfortable feeling that I had not got all the messages from the child as she was attached to so many tubes and I could not adequately convey my emotions with the mask on my face.

Case 4.3

A referral from paediatrics was made for a 5-year-old girl who was admitted with organophosphorus poisoning. The child was in the intensive care unit for a day but had recovered and was now in the general ward. Her older sister who was nine years old was still in the ICU on ventilator while her younger sister who was three years old was evaluated in the ward and found to have very minimal traces. The father was the only earning member and there was no other family support. He reported that he had brought the poison to use for the plants and had left it in their garden. He claimed the children drank it without knowing what the bottle contained. At the first contact, this child had told the paediatric resident that her father had given them the poison to drink but after that, had vehemently denied it.

In Case 4.3, as the child was not forthcoming when spoken to, I decided to see the child in the play room. Given the fact that the treating team was suspecting that the children had been poisoned, daily sessions were planned. Although she wanted her parents to be in the room in the first session, she settled down and, from the second session onwards, she would come by herself. Due to the older sister's condition, the family stayed in the ward without asking for discharge. She would not talk much but played with the available toys, especially the animals and the sand pit. She only started verbalizing around the seventh session. She suddenly said “My father gave that to us to drink”. When I started to probe, she did not elaborate further.

We then planned family sessions and included the grandparents after obtaining consent from the parents. They denied any marital of family issues in this family but agreed to take the children to the village and take care of them there for the rest of the academic year. We could not ascertain the manner in which the children had consumed poison.

The ethical issue here was—what do I do with the information given by the child? After obtaining consent from the child, I discussed this with my medical colleagues and we all decided that protection of the child was the primary goal. The parents would be curious to know what the child had told me and I had to keep this information confidential as the child did not want me to discuss this with her parents. The only way around this was to bring the grandparents into the picture and involve them in the care of the children. The child would want to sit on my lap often but I had to resist the urge to offer comfort and would ask child to sit in a chair close by and talk to me.

Case 4.4
A 12-year-old boy was referred from paediatric nephrology with complaints that he was not cooperative for the treatment regimen which included thrice a week dialysis. He would also not take the oral medication as advised. Due to this, his creatinine did not come down at all. Any plans for transplant had to be stalled due to his unstable physical condition. The child appeared listless and ‘depressed’.

The first plan was to draw up a contract wherein he would come for regular dialysis and attend sessions with me. He honoured the contract and the next four to five weeks, I managed to have about 12 sessions with him. During the individual sessions with the child, he expressed distress at the diagnosis and the chronic nature of the treatment and, more importantly, the costs involved. He reported hearing discussions at home as his parents discussed the source of money for his regular dialysis. He, in one session, broke down and confided that the reason he did not comply with the treatment was because he had lost the will to live and his treatment was causing the family especially his brother and sister to forego basic needs.

The ethical issue I was faced with was ‘what do I do with this information the child has confided in me?’ When I asked him, he refused to let me discuss this with his parents. I had to respect this and continued to speak with the child about the cognitive errors he may be making in deciding that he was a burden to his family. Over the next few sessions, we discussed other ways of looking at the situation and came up with the idea that we would have a family session wherein this would be brought up and the family’s views obtained. His mother broke down when she heard the child relate his anxieties and fears and told him he would never be a burden to them and they would do anything to keep him healthy and happy. His siblings were also involved in this meeting and that proved really therapeutic for the child. Arrangements for financial help for the treatment were made using the medico-social work department's contacts. Following this, the child has been compliant with the treatment and has since also had a transplant.

Case 4.5
A 14-year-old boy was referred from the Department of Paediatric Surgery with complaints of repeated injuries needing sutures. This was the fourth time he had needed sutures and the doctor suspected abuse. On speaking to the child and mother together, both denied any problems at home and school. However, on speaking to his mother alone, it transpired that his father had been hitting the child for not following instructions and doing poorly in studies. His father had reportedly come up the hard way and had done his PhD as a scholarship student. He wanted his son to learn to appreciate all that he had and focus on his studies. The child was not forthcoming in the initial session.

In Case 4.5, during the next few sessions, both the mother and the child opened up and spoke about the difficulties at home. I then evaluated the child for his academic difficulties and found him to have significant mixed learning difficulties along with attention deficit hyperactivity disorder (ADHD). He was started on medication and plans for his academics were also made. His father was also counselled and both the mother and the child reported subsequently that his father no longer hit him. The child was also keen on knowing whether I would reply if he sent me messages over the phone or on Facebook.

The ethical issues faced here were—does this constitute a medico-legal case? In a strict sense, any suspicion of abuse needs to be reported. However, since the father had desisted from hitting the child after being counselled, I decided against this. The other issue was about recording confidential information in his hospital folder. I decided to keep the notes of my sessions with the child and mother separate from the general folder which could be accessed by all the departments. The child's request for access to the therapist by phone or Facebook is definitely a challenge that I may face in the future.

Case 4.6
A 10-year-old boy whose parents were medical professionals and worked in a hospital nearby was brought to the hospital with pain in the legs and not being able to walk for the prior 15 days. He had been refusing to go to school as he found it difficult to go due to this pain. The doctors in Orthopaedics had investigated the child in detail and found no medical problems. On detailed evaluation, the child revealed that he was being bullied in school and was

very anxious about going to school. He was against my discussing this with his parents or his school as the child who was bullying him had threatened him and also, this child's mother was a teacher in the same school.

For Case 4.6, the ethical issues were—how could this child be helped without disclosing the bullying to his parents and to the school? If the child did not give permission to discuss this, could I speak to the parents nevertheless? Was the child old enough to not give permission? Does confidentiality take a backseat when a child's wellbeing is concerned?

The issues became further complicated to resolve as his parents also wanted to be seen after working hours, wanted to know if the therapist would visit their home or would see them in the therapist's home as the child may feel most comfortable. They also did not want to come to the hospital in case they were recognized by other colleagues at the hospital.

Finally, I decided to talk to the child and emphasize the need to protect him from the bully and how, in the long run, it is necessary to learn the skills to deal with such children. After several weekly sessions spread over 8 weeks, during which time the child did not attend school, the child agreed to allow me to talk to his parents in his presence. After this discussion, I also made some boundaries clear to the parents—like the venue and time of meetings—though I did agree to have my sessions with them on a day other than the usual clinic days. These seemed to help and the parents were more cooperative with the plan for the discussion with the school which really helped to clear all the problems the child reported. The child was involved in all the discussions and this helped him realize the importance of confiding.

This ethical issue is an important one—we constantly have to deal with colleagues, family members as well as other professionals who may have a personal connection with the therapist.

4.4 Training Issues

In 2009, the American Academy of Child and Adolescent Psychiatry Executive Council revised its code of ethics and instituted a requirement that member clinicians obtain, and attest to, at least one hour of ethics training every two years. In India, although guidelines are available for ethical clinical practice in almost all specialties, the monitoring process is not in place.

Although most syllabi for clinical psychology M.Sc. and M.Phil. programmes do include one module on legal and ethical aspects of mental health, not too much attention is paid to practical experience in these important components of a therapeutic alliance.

It would be imperative, therefore, as part of a psychotherapy training programme, one or two cases per year are discussed from the ethical and legal standpoint. It is also necessary to teach trainees how to write legal reports and also document ethical issues. These will become only more frequent in the coming years.

In addition, sometimes trainees and sometimes even experienced clinicians are called upon to liaise on a case where there are many medical complications that are not in the field of expertise of the mental health professional. Although the easiest solution is to withdraw claiming lack of expertise, the challenge really is how to train students and professionals alike to be an active member of a multidisciplinary team where a mental health professional will bring her/his unique strengths to an issue that requires more heads than one to handle. This issue is going to become increasingly relevant in the coming years.

The suggestions are basically twofold:

1. Be honest with the treating team that one is feeling not very confident about dealing with the complex medical issues that are involved and ask to be included in discussions so that some clarity is brought in.
2. Discuss with other mental health professionals available to make a plan. It will also be a good idea to read about precedents. Formulating a strategy, taking it to the team, and willingness to make it work will definitely be supported.

4.5 Ethics in Child Psychotherapy: The Way Forward

The idea of psychotherapy itself is a luxury in India that is not affordable for most families. Discussing ethics in psychotherapy seems a great leap that we are not ready for yet. There are a few guidelines but trainees in clinical psychology are not exposed to these guidelines during their training and hence, this is not inculcated into their thinking and action while working with children and their families. It is now left to individual therapists to formulate their own brand of ethics in their practice.

In the medical context, failure on the part of the mental health professional to emphasize confidentiality and need for sensitivity in handling ethical issues to their medical counterparts is one of the reasons for this not being adhered to in routine clinical practice. Having said that, the medical undergraduate training in India is coming of age and, in medical colleges that follow the Medical Council of India guidelines, ethics in medical practice is part of the training programme. There is, therefore, definite hope that future medical undergraduates will at least have exposure to thinking of ethics in their medical practice.

Boundary issues are more of an issue in child psychotherapy as therapists and children too feel some physical contact is acceptable—like patting on the back, ruffling the hair or, with very young children, seating them on the lap as a gesture of affection. However, in the era of boundary violations and legal implications thereof, therapists will do well to monitor their behaviour. The yardstick that most experts agree on, and is a fair one to use, is the level of comfort or discomfort the child displays when the therapist does touch her/him.

The other relevant issue is to answer the question—who is ultimately our client? The parents who have sought treatment and who pay the fees and control the visits, or the child herself/himself? Many a time, the answer to this question is unclear as the goals of the therapy are formed between the parents and the therapist for the child. Obviously, in many cases, the parents need to serve as co-therapists for the therapy to work and, hence, the dilemma.

Many a time, in routine clinical practice, child mental health professionals are called upon to make ethical as well as moral judgements. This will, without doubt, be influenced by the therapist's own biases and prejudices. It is imperative, therefore, that therapists take an objective view as much as possible and always do what is in the best interests of the child, who is and always should be, the primary client (Goldsmith and Joshi 2012).

The way forward is to design guidelines that are relevant to South Asian culture and the way psychotherapy is practised in India. Self-regulation and constant monitoring of one's work and, at the same time, discussing with colleagues and seniors about any dilemmas will be crucial. Awareness of the legal and ethical frameworks within which psychotherapy should be practised is a strong recommendation. Training programmes would also do well to include modules on ethics in psychotherapy practice with special reference to the South Asian context.

References

American Academy of Child and Adolescent Psychiatry. (2009): *Code of Ethics*. Retrieved March 4, 2016 from https://www.aacap.org/App_Themes/AACAP/docs/about_us/transparency_portal/aacap_code_of_ethics_2012.pdf

American Psychological Association. (1990). *Guidelines for providers of psychological services to ethnic, linguistic and culturally diverse populations*. Washington, DC: American Psychological Association.

Ascherman, L. I., & Rubin, S. (2008). Current ethical issues in child and adolescent psychotherapy. *Child and Adolescent Psychiatric Clinics of North America., 17*, 21–35.

Belitz, J., & Bailey, R. A. (2009). Clinical ethics for the treatment of children and adolescents: A guide for general psychiatrists. *Psychiatric Clinics of North America, 32*(2), 243–257.

Goldsmith, M., & Joshi, S. V. (2012). Ethical considerations in child and adolescent psychiatry. *FOCUS, 10*(3), 315–322.

Kassaw, K., & Gabbard, G. O. (2002). The ethics of email communication in psychiatry. *Psychiatric Clinics of North America, 25*, 665–674.

Krener, P. (1995). Ethical issues in pediatric consultation-liaison. *Child and Adolescent Psychiatric Clinics of North America, 4*(4), 723–745.

Schetky, D. H. (1995a). Boundaries in child and adolescent psychiatry. *Child and Adolescent Psychiatric Clinics of North America, 4*(4), 769–778.

Schetky, D. H. (1995). Preface. *Child and Adolescent Psychiatric Clinics of North America, 4*(4), XIII–XXIV.

Wright, W. T., & Roberts, L. W. (2009). A basic decision making approach to common ethical issues in consultation liaison psychiatry. *Psychiatric Clinics of North America, 32*(2), 315–325.

Chapter 5
Classrooms and Counsellor Ethics: A Daily Balancing Act

Neena David

Abstract Schools are being acknowledged as important sites for the delivery of mental health interventions. The dynamic and complex layers that constitute school systems create a unique setting for counsellors working in schools. Counsellors should possess a working knowledge of these nested frameworks when they begin to explore ethical issues specific to school counselling. While counsellors in school settings consider the child as the primary client, they are also required to work with multiple stakeholders who often have differing interests. A counsellor would need to recognize that in navigating through ethical dilemmas, the best interests of the child have to be balanced by appreciating the interconnectedness of collaborative relationships with parents and teachers. The chapter addresses the school counsellor in the Indian context and acknowledges the challenges they are confronted with ranging from the lack of adequate professional training, and meaningful national child mental health policies to a lack of clarity in defining their roles. It examines frequently encountered ethical issues in a school context from a practitioner's perspective. Drawing from established international codes of ethical practice, it encourages counsellors to use ethical decision-making models and processes. The chapter concludes with a core set of ethical considerations that provide a basic framework that would support ethical and reflective practice.

Keywords School counsellors · Ethical dilemmas · Counsellor role · India

Case 1

R, a 10th grade student walks into the counselling room looking visibly upset. This is her first visit and she looks around the room anxiously asking if anyone standing outside the room can hear what she has to say. On being reassured, she proceeds to tell the counsellor that she has been getting into

N. David (✉)
Clinical Psychologist, Counselling Services, Mallya Aditi International School,
Near NIPCCD, Yelahanka New Town, Bengaluru 560106, India
e-mail: neenajd@gmail.com

P. Bhola and A. Raguram (eds.), *Ethical Issues in Counselling and Psychotherapy Practice*, DOI 10.1007/978-981-10-1808-4_5

trouble over the past few weeks with Mrs. T: a teacher known for her abrasive approach with students. Mrs. T sternly conveyed her concern to R about the poor quality of classroom participation and tests. R describes the teacher as being overly harsh and frequently unfairly picking on her, an observation that even her friends apparently agree with. R feels that nothing she does in class will ever meet with Mrs. T's approval. She also reveals that she does not want to share her concerns with her parents as her father had been recently been laid off from his work and arguments about finances between her parents have increased. R recognizes that she will be appearing for a public board examination and that her teacher could give her a low internal grade if she chooses to complain. R reported feeling increasingly upset and anxious. The counsellor is also privy to information that while issues specific to Mrs. T's classroom interactions had been raised in the past, the management appeared to focus on the excellent grades her students received in their public exams.

Case 2
Mrs. B, a parent of a 3rd grade child, requests that her child D be seen by the counsellor as the father had recently announced that he was walking out of the marriage. The abruptness of the announcement had left the family members shocked with Mrs. B insisting that she would not give her husband a divorce and Mr. B claiming that he was going to initiate divorce proceedings. Both parents agreed that D would benefit from seeing the counsellor who would address his emotional concerns. The counsellor establishes a good rapport with the child and keeps both parents informed of progress in the sessions. Mrs. B asks if the counsellor can recommend to her husband that he spend more time with the family and participate in marital counselling for the sake of their child. Mrs. B feels that if this suggestion is to come from the school counsellor, the husband would be more receptive. She follows this up with getting her personal counsellor to speak to the school counsellor about the importance of convincing the husband of the need to engage in marital therapy and to not consider divorce for the sake of their son.

Case 3
During a coffee break, Mrs. P, an elementary school teacher greets the counsellor, and enquires about a particular child who is not in her class: "I saw L coming into your counselling room yesterday; is everything all right with her? I taught her a few years back and she was such a model student. Does her mother know that her child is coming to see you? From my interactions with her mother, I know that she's very ambitious for her children and may not be all right with the fact her daughter was seen by the counsellor. I'm sure L is strong enough to sort out whatever is troubling her—what do you think?"

Case 4
A 16-year-old student working through issues of past sexual abuse and occasional lying behaviours that had got her into trouble with her friends, reveals to the counsellor that she has over the past few weeks begun to help

herself to her friend's parent's liquor cabinet. Both her fairly conservative parents work late hours and the student admits that she finds it fairly simple to access the liquor from her friend's house storage and be able to deal with the evidence before they returned from work. The student expresses concern over this trend but specifically asks the counsellor not to involve her parents and that she would sort it out on her own eventually. The counsellor and the student are aware of the school's zero tolerance policy towards underage drinking.

The potential of school-based delivery systems to dramatically improve access to and positive impact of mental health prevention and intervention services cannot be underestimated (Hoagwood and Erwin 1997).

5.1 School Counselling: Context Matters

Research on preventive and economically feasible mental health models increasingly recognizes the importance of school contexts as sites for intervention (Masten 2003). The National Mental Health Policy document (1982) states under Sect. 7, Education Sector's Role in Mental Health Care: "Social, behavioural and learning problems are manifesting themselves in schools. Addition of mental health inputs in school health programmes is likely to play a major role in their amelioration".

Current research evidence builds a solid case for the provision of school-based services for children's mental health; however, as seasoned school counsellors will attest, the ordered, clinical and logical worlds in which research is generated is often divorced from the intense buzz and layered complexities of classroom dynamics. Mental health practitioners in schools are challenged by the ground reality that a fundamental disconnect exists between educators and mental health practitioners (Ringeisen et al. 2003). Schools, as Adelman and Taylor (1999) wryly remark, "are not in the mental health business, their mandate is to educate". Hence it is not uncommon to find that most mainstream schools view with apprehension and a healthy dose of suspicion any activity or professional not directly related to instruction as taking resources away from their core objective of teaching. The Indian education system is predominantly performance oriented. The need to consider social and emotional aspects of a child, its role and significance in learning and teaching is only beginning to be acknowledged. The translation of that acknowledgement into changes that result in informed and progressive school philosophies and classroom pedagogy is at a nascent stage across a few schools in India.

A senior colleague once remarked, "What is so difficult about counselling? You just have to hold their (students') hands, hear them out and tell them that everything will be alright. You don't even have the task of completing portions or correcting papers!" It is into this climate that a school counsellor in India typically enters.

Huey (1986) observes that the single characteristic distinguishing school counsellors from other mental health professionals, is their work setting. The school context as highlighted by Ringeisen et al. (2003) is a dynamic and complex system continually interacting and changing at many levels. Recognizing the uniqueness of working in schools as a mental health professional provides a framework within which issues pertaining to ethics specific to school counselling can begin to be explored. Working with young individuals and serving different populations simultaneously, the school counsellor is often exhorted to align with Kitchener's (1984) five moral principles regarding ethical decision-making: autonomy (freedom of action and choice), justice (fairness), beneficence (doing good), non-maleficence (avoiding harm) and fidelity (loyalty) and to function always in the best interests of the child. This process can best be described as ambiguous even in contexts where counsellors have clearly defined and written ethical codes and legal parameters that serve to protect the populations they work with.

In the Indian context, the confusion is compounded by variations and irregularities in professional training, inadequate exposure to actual classroom settings, poor access to supervision and the absence of an effective licensing and certifying body that meets international standards. The glaring absence of relevant child mental health laws and a comprehensive policy specific to the promotion of mental health only serves to add to the murky world of being an ethical practitioner in an Indian school setting.

5.2 Defining Roles

A discussion on school counselling ethics takes on relevance and meaning in recognizing the roles a counsellor is expected to perform. The American School Counseling Association (2004), elaborating on the responsibility of the school counsellor states that the role involves helping a child develop effective coping skills, identify personal strengths and assets, recognize and express feelings, as well as to provide a foundation for the child's personal and social growth as he or she progresses through school into adulthood. School counsellors are required to take on multiple roles often in a school climate that has very little understanding of the scope of their training or functioning. In the absence of any clarity regarding their roles, they continue to struggle to define their role, corroborating Paisley and McMahon's (2001) observation that role definition is one of the most considerable challenges facing school counsellors today. It is not unusual for school counsellors to be asked to take on office administrative duties, substitute teacher duties, exam invigilation, manage school public relations, chaperone students to inter-school functions and generally take charge of the extra or co-curriculum programme. School administrators who do not share an appreciation for the presence of an on-campus counsellor, could appear to view the counsellor's role from their perspective and may come to view the counsellor not as a specialist but rather as someone who can fill in the gaps (Burnham and Jackson 2000).

The role of the school counsellor originally involved providing guidance and career information to students. It subsequently also began to focus on providing therapeutic services and psychological assessments. The role has currently expanded to include more consultative aspects. Counsellors may hence be involved in providing inputs on school policies, staff hiring, conflict resolution, student admissions, teaching methodologies and classroom management. They also actively collaborate with teachers in fostering positive mental health by creating and implementing life skills programmes for the student community and relevant issue specific workshops for parents and teachers. The school counsellor may also liaise with and support teachers who may share a counselling relationship with some of their students.

It is imperative that school counsellors develop a good understanding of the school philosophy and how their presence can contribute to the school community at large. Counsellors do need to assume proactive roles in crafting their roles in an organization or risk experiencing a sidelining of their roles and their training. This is undoubtedly a time consuming, continuous and often frustrating process, but one that the school counsellor should be committed to.

5.3 Ethical Challenges

The counsellor takes on a range of roles in a school setting and often works with multiple stakeholders whose interests may differ from one another. While the child has to be considered as the primary client, the school counsellor would need to recognize that allowing for one's counselling framework to be solely guided by what lies in the best interests of the child is rather simplistic and idealistic. Such a position does not do justice to the interconnectedness of relationships in a school context and can begin to gnaw away at the heart of collaborative relationships that the counsellor would need to establish with teachers and parents. The wise African saying: 'It takes a village to raise a child' is perhaps a more pragmatic and humbling approach of viewing counselling practice in schools. It is but inevitable that in trying to maintain a fair balance between different stakeholders and their points of view, the counsellor is challenged by ethical dilemmas practically on a daily basis.

5.3.1 Privacy

At a very basic level the logistics of practicing in a school itself creates ethical challenges. The presence of a school counsellor on the campus signals greater accessibility for the school community in addressing their concerns. Hence there is sometimes a sense of informality and an ad hoc nature to these meetings either between the counsellor and teacher or between student and counsellor. The common spaces teachers and students share—a corridor, a lunch room, a library, a bench outside a classroom may all be settings where a teacher or student approaches the counsellor with, "Do you have a minute? I'm between classes and I really need to talk to you now". Issues pertaining to protecting the privacy of the client's identity are a common ethical challenge. Physical space is at a premium in most schools and trying to find a quiet, secluded corner that can be suitable for a brief conversation is a task. Students and teachers do tend to get curious about who was seen visiting the counsellor's office and may sometimes even offer well-meaning but unsolicited advice or concern.

Case 3 mentioned at the beginning of this chapter is a typical example of the fact that it is virtually impossible for the counsellor to ensure privacy in terms of client identity. The counsellor would need to be careful not to enter into a dialogue with a teacher or a student who makes a query about a client. Non-committal and general responses that seek neither to confirm nor deny the question being asked may also be helpful in dealing with this challenge. Class teachers or section heads may need to be informed in advance that a student will be seeing the counsellor if the session has been scheduled during an instructional period. The counsellor is not required to provide an explanation to the teacher as to the reason for the student seeking counselling.

5.3.2 Informed Consent

Students access counselling services directly or through a teacher or parent referral. Ethical challenges present themselves in either type of referral—a parent may not wish for their child to be seen in counselling, or insist that they can give consent provided they are given a session by session account of counselling, or a child/adolescent may express resentment over the fact that they have to visit the counsellor against their personal wishes. Teachers often perceive that hauling a student to visit the counsellor's office serves as an extension of disciplinary action initiated for the child. Most school counsellors have been at the receiving end of a hassled teacher's litany of complaints of poor behaviour about a student with the expectation that the session with the counsellor will get the offending student to see the error of her/his ways and mend them in time for the next class. In a setting and culture that places high value on respecting elders and obeying figures of authority, does the child truly have a space to refuse consent for counselling?

Adult clients opting for counselling are required to provide their informed consent for the period of counselling. In working with minors the concept of informed consent is a tricky area with some practitioners arguing that entering into counselling is a contractual relationship and a child's minor status does not make him or her eligible to provide such consent. The British Psychological Society (BPS 2009), however, in their Code of Ethics and Conduct on the matter of informed consent, states that a counsellor or psychologist should ensure that clients, particularly children and vulnerable adults, are given ample opportunity to understand the nature, purpose, and anticipated consequences of any professional services or research participation, so that they may give informed consent to the extent that their capabilities allow.

If the parent refuses consent for counselling for their child, the Gillick Principle as mentioned by the British Association for Counseling and Psychotherapy (BACP 2006) Ethical Framework for Good Practice in Counselling and Psychotherapy could be used as a reference point for counsellors in India. This states that, "parental right yields to the child's right to make his own decisions when he reaches a sufficient understanding and intelligence to be capable of making up his own mind on the matter requiring decision".

The counsellor would definitely need to interpret this in a developmentally appropriate manner considering not just the chronological age of the child but the maturity level, the cultural and specific context, circumstances of the case and any special considerations if the child has a diagnosed condition such as an attention or learning disorder.

5.3.3 *Client Identity*

It is necessary for counsellors to remind themselves frequently of who their primary client is—the child in a school context especially at times when they are asked to extend their role to include other members. This has to be balanced with maintaining and nurturing a good working relationship with parents and teachers. Case 2 highlights such an issue where a counsellor may be asked to align with a parent typically when the parents are experiencing long standing marital discord or have initiated divorce proceedings. The counsellor was able to resolve this by reminding the mother that her child and his functioning in school was the focus of intervention. While it was recognized that she and her husband were going through their personal differences, there were limitations in a school counsellor's role and being involved in marital therapy was one such example. The mother was also encouraged to talk this over with her counsellor and to address ways in which she could share a more open communication with her husband. The need for the counsellor to stay impartial and be perceived as fair to both parents was reiterated. With parents going through custody battles for their children, the school counsellor needs to be vigilant in protecting the child's identity as the focus of school-based interventions and in responding fairly to both parents.

5.3.4 *Confidentiality*

The most common dilemmas that school counsellors are challenged by are issues related to confidentiality. Confidentiality is described as "nearly a constant struggle" in school counselling (Phillips 2007). The counsellor needs to respect the students' right to confidentiality and weigh it against what teachers and parents need to know. Under the codes of good ethical practice students do have a right to confidentiality, but this is not absolute or binding and at present there is no clarity in

India, regarding legal guidelines or laws that address issues of confidentiality vis-à-vis children in a school counselling context. In deliberating over confidentiality issues with students and parents, counsellors should factor in that minors are unable to make decisions on their own behalf and are "a group of individuals with few responsibilities, many restrictions, and a complex legal status that maintains a dependency on adults for privilege and access to resources" (Sanger and Willemsen 1991). Minors are dependent on their parents and carers yet they do also demonstrate a progressive movement towards wanting to assert their independence and individuality. Juggling these competing aspects is a truly challenging aspect of the counsellor's daily workload.

Students entrust the counsellor with personal information sometimes with the specific request that their parents not be told that they are seeking counselling services. A crucial component of the counselling relationship is trust. Hence if a counsellor were to communicate information about the student to a parent or a teacher without the permission or knowledge of the student then it does violate the students' trust in the counsellor. It could result in the student withholding any further personal information during the counselling sessions and possibly not receive the required support. On the other hand, a parent has the right to be involved if they are aware that their child may be experiencing emotional or psychological concerns.

School counsellors usually let their student clients know during the first session that there are limitations to confidentiality; they would need to inform their parents if the counsellor felt there was the possibility of harm or danger either to the student client or to others. Situations where the student has spoken of suicidal ideation may allow for a standard counsellor response protocol, however the dilemmas and pondering over possible ethically correct responses occur when the student reveals information about instances such as possession and use of illegal substances and alcohol either by them, their siblings or friends, a possible pregnancy, psychiatric illness in a parent and subsequent child neglect, reporting of an unsound educational practice such as unfair assessment concerns or subject competence about a particular colleague.

Case 1 vignette demonstrated the nested nature of the student's concerns, her rising anxiety levels veering on depression in coping with a harsh teacher sandwiched by her decision not to confide in her parents and the assumed possibility that involving the teacher would result in her being marked down in a board exam. Should the counsellor bring up the issue of her colleague's teaching style with the principal; would it help if the parents recognized that their behaviours or including her in conversations about their troubled finances was affecting her significantly; or would it be wise to involve the teacher directly but in a manner where the teacher would not feel blamed?

In Case 4, the counsellor is aware that in the background of child abuse that the student has shared, issues related to trusting an adult are going to be central to the counselling relationship. Does the counsellor believe in the student's abilities to put an effective stop to her drinking on her own? What evidence does she have from the student's life history that would give her this sense of confidence? Do the parents not have a right to know? What if the parents found out inadvertently from the daughter; would not they feel that if they had been informed earlier they could have made a difference to her life? If the student was engaging in an illegal behaviour such as underage drinking, should the school authorities know or would withholding this piece of information about the student be perceived as the counsellor colluding with the student against them?

5.3.5 Teachers as Counsellors

In a country of 1.2 billion citizens and counting, the statistics of the number of trained mental health professionals in relation to the population that require their services are dismal. The need to look at alternative/community-based options for early identification and intervention of emotional and/or behavioural conditions in children has increasingly focused on schools as viable sites. Sensitization programmes on mental health in schools (Kapur 1998) alert teachers to early warning signs of disturbance and working on basic counselling skills with children and adolescents.

Teacher–student interactions over time allow for counselling opportunities to arise in an organic process. Teachers can sometimes be drawn into establishing a counselling role with a student who chooses to confide in them. This may occur in schools where there is no designated counsellor as well as in schools where there may be a professional on campus (Miller et al. 1993). Teachers enter into counselling situations out of a need to address the stress that the student may be experiencing. The ethics of teachers as counsellors is a grey zone and one that does not get addressed often enough by school counsellors. It is fairly common for the school counsellor to be approached by a teacher who mentions that a student has confided in her about a personal issue but does not want to meet with the counsellor to address it. The counsellor takes on a secondary role—that of a proxy counsellor and supervisor, supporting the teacher to the best possible extent.

Teachers are required to take on pastoral duties in their classrooms in interacting with their students; these, however, need to be viewed differently from a teacher who engages in in-depth counselling with students. The ethical issues that the teacher practitioner has to contend with are recognizing the boundaries of their role in terms of acknowledging a power differential in the relationship, role conflicts between a traditional teacher role versus that of a non-judgemental and supportive counsellor, training, counselling competence, availability and time. The counsellor should also be quick to spot early signs of the teacher feeling emotionally burdened with this role and when it becomes imperative to make a professional referral in a manner that allows for the student to continue experiencing a sense of support and caring from her teacher.

5.4 Resolving Dilemmas

Working towards resolving an ethical dilemma can sometimes be a draining experience for the school counsellor emotionally and professionally. It does require a certain level of ethical awareness and competence. It should be stated that school counsellors need to operate with a flexible orientation and make decisions based on the unique features of each case and actively consider their cultural contexts and school settings. Timm (1999) in his study on school counsellors reported that in handling ethical dilemmas, a majority of counsellors preferred to use peer consultation. Deliberating over the risks and benefits of their actions and using a systematic decision-making model were other ways through which they engaged with resolving ethical dilemmas.

The focus of literature appears to recognize that student confidentiality is the 'primary ethical dilemma' for school counsellors (Bodenhorn 2006). Boderhorn's exploratory study indicated that counsellors in elementary and middle school "experience a higher rate of challenge than their high school colleagues in facing some dilemmas, namely dual relationships with faculty, student confidentiality and parental rights". A possible explanation for this finding would be that counsellors serving this age group work more closely with teachers and parents and hence have a higher frequency of interactions with them. The Ethical Standards for School Counsellors drafted by the American School Counselor Association (ASCA 1998) and the Code of Ethics and Standards of Practice drafted by the American Counseling Association (ACA 1995) both specify that regardless of client's age, counsellors can breach confidentiality if clients "engage in behaviour that presents clear and imminent danger to themselves or others". What constitutes 'danger' needs to be interpreted in context of the child's age, coping mechanisms, shared values and family and school support systems.

The ASCA in its document on Ethical Standards for School Counsellors(2010) recommends that, when faced with any ethical dilemma, school counsellors, "use an ethical decision-making model such as Solutions to Ethical Problems in Schools (STEPS)" (Stone 2005): This model consisting of nine steps suggests that counsellors proceed through specific steps that get them to consider the emotional influences of a problem, the chronological and developmental appropriateness of the client and the solution, the setting, the child's and parent's rights, possible courses of action and their consequences and consultation with a supervisor or experienced peers prior to implementing the decision.

1. Define the problem emotionally and intellectually.
2. Apply the ASCA and ACA Code of Ethics and the law.
3. Consider the chronological and developmental levels.
4. Consider the setting, parental rights, and minors rights.
5. Apply the moral principles (Kitchener 1984).
6. Determine the potential courses of action and their consequences.

7. Evaluate the selected course of action.
8. Consult.
9. Implement the course of action.

While some of this specifically relates to counsellors working in American schools, it does allow for a reference point for school counsellors in India. Mandating licensing requirements to practice as counsellors in Indian schools, creating opportunities for research and continuing education and establishing a code of ethics is the crying need of the hour. In the absence of any mandatory requirements, a lack of clarity about the role of the school counsellor and insufficient access to supervisors or experienced peers it falls upon the individual practitioner to actively develop a reflective practice and carry out their responsibilities with respect for their clients, competence and integrity.

5.5 Ethical Practice: Considerations

Drawing on the ASCA and BACP Code of Ethics, this chapter concludes with a core set of ethical considerations that school counsellors may find relevant in their practice. These ethical considerations do not presume to be comprehensive enough to cater to the range of roles that a counsellor is required to handle in a school; they hope to provide instead a basic framework that would encourage ethical practice.

5.5.1 Responsibilities to Students

Counsellors should treat all students with dignity and respect. Students come in from a range of different belief and value systems and cultural contexts which need to be acknowledged and respected. The counsellor should be mindful of not imposing their values or what constitutes right/moral behaviour on the student.

Encourage and support the student's need to move towards personal and social growth.

Teachers and parents may have their specific views of the student that may well be accurate but the counsellor is required to maintain an objective attitude and an open mind in listening to the student.

Explain the limits of confidentiality to the student in age-appropriate language. Inform parents/care givers if there is danger to the child's or other's life.

Maintain an appropriate professional distance during and between sessions and also on social networking sites.

Necessary referrals to competent external authorities should be made when the counsellor recognizes that the student's range of difficulties is not what they have sufficient training for or that the student may benefit from multidisciplinary inputs.

If the student requires psychological and/or cognitive assessments which would provide useful information in providing appropriate interventions, the counsellor should use relevant and appropriate tests that are current and—have training for in terms of administration and interpretation. Test information should be communicated in a manner that can be understood by the student and the parent.

The counsellor should exercise care in storing confidential student records such as session notes, test reports and profiles.

When working with students in groups, recognize that confidentiality, how students respond to it and whether they choose to honour it cannot be guaranteed. When engaging in group work with younger children, the counsellor should seek parental informed consent.

The counsellor should be conscious of not aligning with the student—or taking sides against the school or parent in an effort to be popular with students or wanting to be appreciated by the student.

5.5.2 Responsibilities to School

The counsellor should have a good understanding of the school's vision and philosophy.

Each school follows its unique set of processes in regulating the events of the day; some of these may be unwritten codes that the counsellor would need to be aware of.

Defining the role of the counsellor, boundaries regarding what information can be shared and sensitizing school administration and faculty to the functioning of counsellors on campus is an ongoing task that the counsellor needs to engage in at formal levels, such as workshops, staff meetings, sending out flyers or emails and at informal levels through regular conversations.

Most schools are strongly hierarchical in nature and they tend to have clear and fixed lines of command. The counsellor would need to be aware of and respectful of these.

The counsellor should stay informed of existing and new school and affiliated exam board policies and how it impacts school and student functioning.

The counsellor should strive to develop collaborative relationships with school administration and faculty and recognize the importance of their roles.

Counsellors should be appreciative of the teachers' workload in context of the complexities of classroom dynamics and in addressing issues of classroom or specific student management—recognize teachers as collaborators and not assume a prescriptive approach with what they think will work in the classroom or benefit a student.

If a student is visiting the counsellor during an instructional period, the counsellor should be prompt in informing the teacher concerned and if possible avoid seeing a student when the teacher has announced a test or is introducing a new topic to the class.

The counsellor should conduct themselves with dignity and maintain professional relationships with school faculty.

5.5.3 Responsibilities to Parents

In working with minors, counsellors would need to remember that legal rights of the parents supersede ethical concerns of the counsellor.

The counsellor would need to establish collaborative relationships with the parents and share accurate and objective information with them that would be in the best interests of the child.

Counsellors should treat all parents with dignity and respect, recognizing that they come from a range of cultural backgrounds and value systems.

Parents should be invited to meetings where counsellors can provide information on the counselling process in an effort to demystify it and clarify parent queries.

5.5.4 Responsibilities to Self

The counsellor should identify and consult with an experienced peer or supervisor regularly as this would benefit their reflective practice and also allow them to feel supported in their sometimes difficult moments.

Counsellors should recognize their limitations in working with different stakeholders in terms of their training and experience and take responsibility for their actions. They should seek to enhance their professional competence by participating in conferences, attending training workshops and staying engaged with latest research in the field.

Counsellors do experience stress which may impact on their competence levels. It is hence important that they attend to their physical and emotional health and wellbeing.

References

Adelman, H. S., & Taylor, L. (1999). Mental health in schools and system restructuring. *Clinical Psychology Review, 19*(2), 137–163.

American Counseling Association. (1995). *Code of ethics and standards of practice*. VA: Alexandria.

American School Counselor Association. (2004). Ethical standards for school counselors. Retrieved October 9, 2006, from http://www.schoolcounselor.org/content.asp?contentid=173

American School Counselor Association. (2010). Ethical standards for school counselors. Retrieved on March 5, 2016, from http://www.schoolcounselor.org/asca/media/asca/Resource%20Center/Legal%20and%20Ethical%20Issues/Sample%20Documents/EthicalStandards2010.pdf

Association, American School Counselor. (1998). *Ethical standards for school counselors*. VA: Alexandria.

Bodenhorn, N. (2006). Exploratory study of common and challenging ethical dilemmas experienced by professional school counselors. *Professional School Counselling, 10*(2), 195–202.

British Association of Counselling and Psychotherapy. (2006). *Good practice guidance for counselling in schools* (4th ed.), British Association of Counselling and Psychotherapy.

British Psychological Society. (2009). Code of Ethics and Conduct. British Psychological Society. Retrieved from http://www.bps.org.uk/system/files/documents/code_of_ethics_and_conduct.pdf

Burnham, J. J., & Jackson, C. M. (2000). School counsellor roles: Discrepancies between actual practice and existing models. *Professional School Counseling, 4*(1), 41–49.

Director General of Health Services. (1982). *National mental health programme for India*. New Delhi: Ministry of Health and Family Welfare.

Hoagwood, K., & Erwin, H. D. (1997). Effectiveness of school based mental health services for children: a 10 year research review. *Journal of Child and Family Studies, 6*(4), 435–451.

Huey, W. C. (1986). Ethical concerns in school counselling. *Journal of Counseling and Development, 64*(5), 321–322.

Jacob-Timm, S. (1999). Ethical dilemmas encountered by members of the national association of school psychologists. *Psychology in the Schools, 36*, 215–217.

Kapur, M. (1998). *Mental Health in Indian Schools*. New Delhi: Sage Publications.

Kitchener, K. S. (1984). Intuition, critical evaluation and ethical principles: The foundation for ethical decisions in counseling psychology. *Counseling Psychologist, 12*(3), 43–55.

Masten, S. A. (2003). Commentary- Developmental psychopathology as a unifying context for mental health and education models, research and practice in schools. *School Psychology Review, 32*(2), 169–173.

Miller, J., Manthei, R., & Gilmore, A. (1993). School Counsellors and guidance networks: roles revisited. *New Zealand Journal of Educational Studies, 28*(2), 105–124.

Paisley, P. O., & McMahon, H. G. (2001). School counseling for the twenty-first century: Challenges and opportunities. *Professional School Counseling, 5*, 106–115.

Phillips, E. (2007). The dilemma of practicing and teaching confidentiality within the school setting. In S. Isreal & C. Lassonde (Eds.), *The ethical educator* (pp. 39–48). New York: Peter Lang Publishing Inc.

Ringeisen, H., Henderson, K., & Hoagwood, K. (2003). Context matters: schools and the research to practice gap in children's mental health. *School Psychology Review, 32*(2), 153–168.

Sanger, C., & Willemsen, E. (1991). Minor changes- Emancipating children in modern times. *Journal of Law Reform, 25*, 239.

Stone, C. (2005). *School counselling principles: Ethics and law*. Alexandria, VA: American School Counselor Association.

Chapter 6
Ethical Issues in Working with Couples and Families

Sobhana H., Priya Saxena and Buli Nag Daimari

Abstract Most mental health practitioners of couple and family therapy agree that engaging couples and families in therapy is a very challenging prospect. Like all therapeutic interactions, the practice of couple and family therapy is also governed by ethical frameworks. In India, there is no statutory body that specifically regulates the practice of couple and family therapy. Hence, most therapists rely on interpretations and adaptations of codes of ethics from other countries. Practitioners also rely on their experiences and consultations with peers and colleagues to deal with situations that pose serious ethical issues. This chapter primarily focuses on ethical issues with regard to responsibility, confidentiality and therapist's values. Other ethical concerns like multicultural competencies, maintaining professional boundaries and legal obligations that are specific to Indian context are also discussed. Each of these issues is explained, followed by case illustrations drawn from the authors' clinical practice. The case illustrations highlight ethical concerns contextualized to the Indian couple and family therapy practice setting.

Keywords Ethical issues · Couple and family therapy · Values · Family therapy in India · Ethics and law

Personal and professional values and ethical standards are vital in any form of therapy. As mental health professionals we have to understand the underlying values of the couple and family system, the context in which we work and our own personal

H. Sobhana (✉) · B.N. Daimari
Department of Psychiatric Social Work, LGB Regional Institute of Mental Health,
Tezpur, Assam 784001, India
e-mail: sobhana@gmail.com

B.N. Daimari
e-mail: bulidaimari@gmail.com

P. Saxena
Department of Clinical Psychology, LGB Regional Institute of Mental Health,
Tezpur 784001, India
e-mail: priyasaxena175@gmail.com

P. Bhola and A. Raguram (eds.), *Ethical Issues in Counselling and Psychotherapy Practice*, DOI 10.1007/978-981-10-1808-4_6

values. Ethical standards provide the guidelines on expressions of values and provide us with the "dos and do nots" of working in a therapeutic context. Therapists face several situations that call for ethical judgements and decision-making to find solutions to dilemmas and conflicting areas (Ables and Brandsma 1977).

Most couple and family therapists will agree that engaging their clients in therapy has many challenges. Every stage of therapy requires making ethical decisions that can impact the course and outcome of therapy. Ethical decision-making requires reflective awareness and sensitivity. Discordant personal and professional values can be a source of ethical dilemmas for practitioners and there are often no easy solutions to these conflicts. In such a scenario, the practitioner must consider the context of the client; their rights, professional obligations and integrity. Gambrill (2008), Strom-Gottfried (2008) recommend considering the following factors:

- The couples'/families' interests, rights and values.
- The interests and rights of others involved in the situation.
- The professional code of ethics and how this relates to the situation.
- Personal values and ethical stance, practice setting, policies and procedures that relate to the situation.
- The legal and licensing regulations and implication of each intervention decision.

In the West, professional organizations like the American Association of Marital and Family Therapists (AAMFT 2012) clearly lay down the ethical code of conduct for marital and family therapists, which are updated from time to time. The AAMFT code consists of the following areas: responsibility to clients, confidentiality, professional competence and integrity, responsibility to students and supervisors, responsibility to research participants, responsibility to profession, financial arrangements and advertising (Goldberg and Goldberg 2008).

Unlike other countries, in India, in the absence of a statutory body to regulate the practice of couple and family therapy, practitioners tend to rely on interpretations and adaptations of code of ethics from other countries as well as on training, experiences with clients and support from peers and colleagues. The recent draft version of the updated ethics guidelines of the Indian Association of Clinical Psychologists (IACP 2015) has a section focused on working with couples and families. Some of the pertinent areas of ethics to be kept in mind while working with families and couples as seen from ethics codes in other countries and in India are delineated below.

6.1 Competence of Therapist

The therapist must have the necessary qualifications, training and skill. The guidelines also delineate the context in which the therapist's competence needs to be ascertained; for example, while working with couples/families, understanding of psychological issues in relation to age, life cycle stage of the couple/family, sexual orientation of the couple and the cultural upbringing of the family members. A competent therapist should endeavour to update knowledge of current developments in the field and show sensitivity towards psychosocial nuances in the process of couple/family therapy.

6.2 Responsibilities of the Therapist

Practitioners are accountable towards their clients and should refrain from taking steps that can harm clients' wellbeing. This becomes specifically relevant in the context of couple/family therapy as more than one member is defined as the 'client'. Ensuring that the commitment remains fair and just towards all members of the couple/family unit is much more complex. Inherent in the principle of responsibility is the need to take informed consent from the client and the responsibility to lay down clearly the nature of the relationship and flow of information when a third party is involved. These issues present unique challenges in couple/family therapy.

6.3 Therapeutic Contract

The therapeutic contract involves clarity about the responsibilities of clients and therapist, confidentiality of shared information and the structure of therapeutic sessions. The IACP has specified the exceptions to the confidentiality clause when there is sufficient evidence of abuse in the family.

6.4 Ethics of Responsibility Towards the 'Client': A Balancing Act

To throw some light on ethical dilemmas and contextual decision-making, a few relevant couple/family therapy case scenarios are illustrated which concern therapist responsibility, confidentiality and informed consent.

Very few mental health institutes in India have full-fledged couple and family therapy units and centres that provide focused training and experience in this area leading to concerns about the adequacy of training and therapist competence. A therapist trained in individual psychotherapy needs to shift from the individual focus to the systems focus while the definition of client has to shift from the individual to the couple or family.

This shift necessitates addressing questions of ‘who’ the client is and where does the therapist’s responsibility lie? Typically, in a mental health setting, an individual with a psychiatric disorder is referred for couple/family therapy when the treating team observes that interpersonal relationships and conflicts are significantly affecting the treatment and recovery. At times, therapy is sought when one member feels the need for help and initiates therapy for the family. In either case, the therapist needs to guard against forging an alliance with one family member at the expense of the others. A case example of a couple referred for therapy would help to illustrate this dilemma (see Case 6.1).

Case 6.1
Mr. AH, 35 years old, had sought treatment for anxiety and depression. A marital therapy referral was made in the context of his preoccupation that, like his first wife, his present spouse too would divorce him over the frequent conflicts they had. Prior to bringing his spouse into the therapy room, the evidently anxious client requested that the therapist avoid questioning her about their marital conflicts. He described her as very temperamental and was sure that this would result in another fight and the dissolution of their marriage. The therapist listened to what Mr. AH had to say, encouraged him to come in for a joint session with his wife and assured him that his concerns would be addressed during the sessions.

In the scenario described in Case 6.1, the spouse who initiated therapy was trying to forge an alliance with the therapist, by approaching the therapist separately and trying to “warn her” of what was inadmissible content in therapy. A beginning therapist, uncertain about his/her competence could easily be swayed by this and skirt the problem area for fear of adversely impacting the alliance with the client. However, this does not mean that the concerns of the spouse, who approached the therapist first, should be disregarded. Typically, couples who seek marital therapy have experienced a lot of distress and carry their own anxieties about the relationship. They also have their own notions about the therapeutic process, often expecting quick changes, provided the therapist ‘explains’ to the other spouse, or “makes the other spouse understand”. Therapist neutrality is an ethical imperative in such situations.

The question as to which member the therapist is responsible to is a constant challenge throughout therapy. A couple/family therapist often finds that one of the members tries to tilt the flow of therapy in his/her favour, by asking questions such as, “Am I not right in expecting this much?” Such questions reflect their implicit assumptions and hope that the therapist shares a similar world view. It is important to communicate to the members that the focus of therapy and the primary target of improvement is the relationship, and not individual demands or expectations of the family members. The fulfilment of individual needs/expectations is in the service of improving the marital/family relationship. The couple/family therapist is a relationship advocate (Margolin 1982), and not a champion for individual needs.

Another common ethical quandary for couple/family therapists relates to whether therapy should be continued when a key family member clearly shows dissent over the continuation of therapy. In situations such as these, the therapist has to decide where his or her responsibility now lies. Does the therapist have to continue therapy with one spouse or insist on the presence of the other spouse also? These issues are discussed in Case 6.2.

Case 6.2

Mrs. JZ was a 39-year-old, high school educated woman from a traditional background. Her 19-year-old son, the eldest of three children, was brought to the hospital for treatment of behaviour problems, exhibited after contracting encephalitis a year earlier. In the course of treatment, the treating team suspected problems in the family and made a family therapy referral. The husband refused to come for therapy, but upon the insistence of the treating team, he reluctantly attended one session of assessment. Throughout the session, he refused to speak or cooperate. The therapist noticed that the wife was reluctant to speak in the husband's presence and wondered if her hesitation was due to her cultural background and upbringing. Sensing this, the therapist had individual sessions as well as sessions with her other children to help them express their concerns as a family. Revelations in the sessions indicated that the husband did not provide adequate financial support, had extramarital involvements and Mrs. JZ could be at risk for sexually transmitted diseases. There were also some instances of the husband's sexually inappropriate behaviour with their girl child.

Case 6.2 brings out several ethical issues for the therapist to consider. When one family member has clearly indicated displeasure and dissent over continuing therapy, is it ethical to repeatedly request him to reconsider? Since it was clear that he would not participate in the therapy, then does the therapist assume the responsibility for the welfare of other family members? Since it was evident that there were several grave issues that had to be addressed, the family therapy unit decided to go ahead with individual therapy to address couple-related problems and empower the wife to deal with the problems on her own. Family therapy with the children and mother focused on strengthening their relationship, redefining their roles and the husband's inappropriate behaviour towards the daughter. The underlying objective of therapy was to empower the mother–children subsystem.

Individual marital therapy (IMT) has received mixed reviews from family and couple therapists (Gurman and Kniskern 1978; Wells and Giannetti 1986).

Recently, Shah and Satyanarayana (2011) demonstrated that systemic individual marital therapy (SIMT) may be suitable in contexts where only one partner is available.

Case 6.3
Mr. K was referred for family assessment and intervention for marital conflict. Assessment indicated that Mr. K had remarried with consent from the first wife, with the aim of having a progeny. The couple agreed that the child would be looked after by both the wives. The second wife had been diagnosed with a mental illness prior to the marriage and it was with the consent of "all members" that the marriage took place. Soon after the husband's second marriage, the first wife expressed strong disapproval of her husband's sexual involvement with the second wife.

The first challenge in working with Case 6.3 was in the definition of the 'couple'. The American Psychological Association (2010) ethics code states that, "When psychologists agree to provide services to several persons who have a relationship (such as spouses, significant others, or parents and children), they take reasonable steps to clarify at the outset (1) which of the individuals are clients/patients and (2) the relationship the psychologist will have with each person". In this complex situation, the first wife and the husband had come together for therapy and there was a strong bond between them. It was important to understand the concerns of the first wife, given the entry of the second wife. It was acknowledged that the decision to remarry was solely for the purpose of procreation. It was equally important to help the couple comprehend the ramifications of their decision of this remarriage. The needs and rights of the second wife also had to be considered here. It was an ethical challenge for the therapist to have the second wife present in the conjoint sessions, especially because the latter was not able to participate fully owing to her psychiatric condition.

In this context, the therapist's responsibility was not only towards the couple who sought therapy but also towards the second wife who was a significant member in this relationship. She was vulnerable because of her mental illness and appropriate treatment was initiated for her. The therapist took an ethical decision to become the voice of the weaker member of the family, in order to equalize the power differences. Through the sessions, the first wife was helped to empathize with the position of the second wife. It was important to help the husband develop commitment towards taking care of the second wife, and provide for her financially in the future also.

Mental health professionals working in the rural or semi-rural settings in the Indian may encounter such marital alliances and the attendant disequilibrium. The ethical challenges cannot easily be answered by professional codes of ethics and therapists need to balance the needs of all family members in the family unit.

6.5 Confidentiality in the Marital Subsystem: Keeping Multiple Secrets

In the absence of a confidential relationship, couples/families may find it difficult disclose personal and sensitive information (Thomlison and Corcoran 2008). Though there are guidelines to help professionals make ethical decisions, they do not clearly state the appropriate behaviour expected from practitioners.

The issue of confidentiality could arise in the following situations. A member may be unwilling or unprepared to reveal certain relevant but sensitive information in the presence of the other family members. A spouse or family member could also attempt to reach the therapist over phone or have individual contact for 'extratherapeutic communications', to privately convey disinterest in therapy or try to further an individual agenda. When shared information is relevant to the marital relationship, this creates a difficult situation for the therapist.

It is essential that in the beginning of therapy, the therapist make his/her position on confidentiality clear. There are three positions to choose from: (a) information given during individual sessions is treated as confidential and not divulged to the spouse but the person may be encouraged to share this information in a conjoint session later on; (b) avoid any separate sessions with either spouse and make sure secrets are discussed openly during the couple session; and (c) have individual session and contacts but clearly inform and take consent from the couple that the therapist would make a decision or judgement call on whether to divulge any information if this is viewed as being beneficial to the couple and to their relationship (Goldberg and Goldberg 2008; Corey et al. 2011).

Case 6.4

Mr. and Mrs. RK were married for 10 years. A couple of years ago they had sought treatment for the wife's use of an injectable opioid analgesic. The treatment had focused on her substance use as well as their relationship issues. In the recent consultation, the couple came with a request for marital therapy after Mr. RK found evidence of a letter written by his wife on the computer, indicating some level of intimacy with the intended recipient. He wanted to know the person's identity and the extent of their relationship but his wife refused to answer his questions. However, she was ready to meet the couple therapist and work to rebuild their marriage since they were having frequent arguments over this matter.

Throughout the assessment sessions, the husband would express his need to know the truth about his wife's affair and his inability to move forward in the relationship without discussing this. During the concurrent session with the wife, she confided that she had two extramarital involvements which involved sexual intimacy, one of which husband was completely unaware of. She had ended both these relationships a while ago and she forbade the therapist from bringing this issue in the conjoint session.

In Case 6.4, the therapist experienced an ethical dilemma over the wife's confidential dilemmas. The therapist considered the ethical implications and took a call on not addressing issues related to wife's first extramarital relationship that ended long ago as this had no immediate relevance to their relationship. When a partner discloses an extramarital involvement (EMI), it is advisable to provide space for individual sessions so that the involved partner can weigh the options in a secure environment (Agrawal and Shah 2009). In the present case, concurrent sessions helped the wife understand husband's anxiety and distress over her EMI. The wife was encouraged to discuss the pros and cons of disclosure as well as concealment of this information. Concurrent sessions with husband helped him realize that his wife has to be comfortable to discuss her involvement. He was helped to understand that his insistence could ruin the chances of reconciliation and he waited till his wife was ready to discuss. The wife waited till she felt confident to disclose details of her EMI. She requested the therapist to facilitate this discussion with the husband. By addressing the concerns of both spouses, the therapist was able to effectively manage the confidentiality issue.

In the process of couple therapy, one spouse could make disclosures that require a careful ethical decision-making process. The assurance of confidentiality enables clients to share personal and intimate experiences (Younggren and Harris 2008), but the exceptions to confidentiality, including information that may be shared without the client's consent, must be spelt out during the informed consent process (Avasthi and Grover 2009; Bennett et al. 2007). Circumstances that involve reporting abuse, protecting lives of clients and intended victims, may warrant sharing of confidential information without consent.

Case 6.5

Mr. and Mrs. IB sought therapy to deal with constant quarrels and domestic violence. This was the third marriage for Mrs. IB and first for Mr. IB. On two occasions, the wife had lodged a complaint against her husband for domestic violence which she subsequently withdrew. A week before consultation, the couple had an altercation which involved physical violence towards the wife. The wife threatened to lodge a complaint of domestic violence against her husband and seriously pursue it this time. During the intake session, she expressed that she wanted to work on the marriage and avoid another divorce. The husband also expressed his commitment to the marriage. The conjoint sessions were full of arguments and so concurrent sessions were held with the husband and wife. In the concurrent sessions, the husband revealed that his participation in therapy was based on legal advice aimed at forestalling the domestic violence complaint and that he planned to initiate divorce proceedings later at the right time. The husband would repeatedly call the therapist and request help in convincing the wife not to press charges.

In Case 6.5, it was clear that the husband had come with a hidden agenda to use therapy to escape legal consequences for his actions and was trying to manipulate the system as well as the spouse. At this juncture, the therapist brought this matter up for discussion in a conjoint session and took the decision to terminate couple therapy.

The practice of family therapy in a patriarchal society, where power and gender inequality is deeply ingrained, presents compelling challenges to maintaining confidentiality of less powerful members in the family. A mother-in-law could insist on knowing information shared by the daughter-in-law; or a father could be persistent on being 'updated' about his son's 'revelations' in individual sessions. The sharing of information or lack of it could result in an adverse event for the person concerned. Often, if the powerful member in the family feels excluded and sidelined, he or she could thwart the progress of therapy and pose emotional or physical threats to the weaker member. Hence, the practitioner should tread carefully and sensitively and make an ethical choice. It is true that a pre-stated therapy policy on confidentiality can, to an extent, help navigate this situation.

Case 6.6

Mrs. M, employed in a private firm, approached the therapist seeking help in negotiating with her parents-in-law on several family matters. Mrs. M's husband was the only child of an ageing couple and was still financially dependent on them for major expenses. His parents had strong opinions about how their grandchildren had to be brought up and disapproved of their daughter-in-law's approach. The couple was not allowed to make independent decisions and the husband was unable to influence his parents to change their stance. The wife could not openly express her emotions in the family sessions in the presence of the in-laws. Separate conjoint couple sessions with her and her husband were initiated.

In Case 6.6, Mrs. M and her husband did not want to relocate away from the husband's parents and sought solutions that would not alter the current family structure. She was aware of her position in the family and did not want to challenge the power hierarchies in the family. The parents-in-law disapproved of therapy but were willing to participate. Although confidentiality issues were discussed at the outset, they continued to insist on knowing what was discussed in the couple sessions. When the therapist refused to divulge this confidential information, they then demanded disclosures from the couple after their sessions.

This posed a great challenge in the smooth conduct of therapy and such situations reflect the cultural perception of a marital system in India. It is not uncommon to find the couples who are structurally nuclear but functionally extended. Most often, cross-generational subsystems like that of the parent–son allegiance takes precedence over the husband–wife relationship. Couple therapists in India often find that there are more than two members in the marital system (Nath and Craig 1999). The process of couple therapy may often pose a threat to the power hierarchies and seek redefinition of the traditional structure of the family. A culturally sensitive and ethical perspective in family therapy accounts for concerns of all members (including the parents-in-law in this case) and empowers members or subsystems in a planned manner (Tamura and Lau 1992; Almeida 1996).

6.6 Therapist's Values: Intersections with Ethical Frameworks in Therapy

Therapist values can influence the process of couple and family therapy and biases may occur when the family's value systems, cultural backgrounds, practices and attitudes are at odds with those of the therapist. Some potential spaces for ethical conflicts include therapist attitudes regarding diverse sexual orientations, divorce, alcohol use, gender roles and equality and traditional societal values. The following case vignettes (Cases 6.7, 6.8) illustrate how therapist personal values can enter the therapy room and call attention to the need for self-awareness (Goldberg and Goldberg 2008).

Case 6.7
A family was referred for therapy to address their frequent conflicts which affected the treatment response of the 26-year-old daughter who had a diagnosis of paranoid schizophrenia. Assessment revealed that the father was dominating, emotionally and sexually abusive to the mother but caring and protective towards his daughter. The trainee therapist, who had very strong views about gender equality, would inadvertently let the wife vent against the husband and would hardly allow the husband to voice his views. After three to four sessions, the husband expressed that he was not being understood, was being singled out in sessions and asked for an older male therapist. During supervision sessions with her supervisor, the trainee therapist became aware of this impasse. Through the supervisory sessions she was made aware of how her gender-related beliefs and values had interfered with her neutrality.

The nature of problems couples or families bring to therapy often trigger expression of therapist's own attitudes and biases. Any discomfort encountered during therapy should either be discussed or handled with colleagues, supervisors or others in the profession.

Closely linked to the attitude and value systems are the moral positions therapists tend to take on what is right and wrong for the couple or family and what is acceptable/unacceptable behaviour. This kind of positioning can jeopardize the therapeutic relationship and defeat the very purpose of therapy.

Case 6.8
A couple in their mid-30s sought therapy for marital issue and arguments over reaching the decision to have a child. In a concurrent session, the husband revealed that he was gay; he had thought things would 'change' after marriage. He was now experiencing significant distress related to problems with sexual intimacy in the marriage.

The therapist of the couple in Case 6.8 was a devout Christian who had strong views on sexuality and sexual orientations. Until the husband disclosed his sexual orientation, the therapist was able to empathize and facilitate therapeutic discussions. After the disclosure, the therapist realized that her religious beliefs were impeding the neutral stance necessary in therapy. In the interests of the welfare of the couple, she decided to refer the couple to another therapist.

While, referral to another therapist is one ethical choice to ensure that we "do no harm", there are contrasting opinions about this approach. Another perspective stresses the need to bracket one's personal values and operate from the professional values of equality and social justice. While adopting such a stance, the therapist can seek outside consultation with experienced therapists who have worked with couples from a sexual minority background (Jackson et al. 2013). Developing competence in working with LGBT (Lesbian Gay Bisexual and Transgender) couples or persons can be a challenge for therapists who have inadequate knowledge of psychosocial trajectories of development and inadequate training (Pachankis and Goldfried 2013).

Therapeutic engagement with same-sex couples means that the therapist must be willing to explore different aspects of the relationship. Research has indicated that that same sex couples respond to therapists' perceived discomfort with exploring sexual issues, by 'protecting' the therapist and limiting disclosure. The therapist may err on the side of being too affirming and fail to explore negative aspects of the couples' relationship, including aspects such as intimate partner violence (Cannon and Buttell 2015; Grove and Blasby 2009).

In the changing urban landscape, another challenging form of marital relationship is open marriages, in which couples espouse the primacy of the marital relationship, but are mutually open to satisfying their emotional, sexual and spiritual needs outside the marriage (Duggal 2014). Therapist belief systems about monogamy could bias the decision making or impair the therapy process. Ethical concerns would also involve awareness about the balance of power between the husband and the wife, and attention to the needs of both partners (Duggal 2014).

6.7 Managing Boundaries in Couple and Family Therapy

The poor educational status, socio-economical and geopolitical situations of the clients often present contexts which are unique to Indian culture. As couple/family therapists working in the North East region of India, therapists often face challenges to the boundaries of their professional roles.

Case 6.9
Mr. Q, a daily wage earner, was seeking treatment for depression in the background of marital discord over their inadequate financial resources. His wife and mother-in-law refused to allow him to stay with the family and threatened him with divorce. Mr. Q was unable to get in touch with the wife and requested the therapist's help in contacting her. The therapist then sought help of a mutual friend who was also a hospital staff to convince the wife to meet the therapist. The wife came in for one session and discontinued after which individual therapy with the husband continued. As a sign of gratitude for her 'efforts', Mr. Q offered a cup of tea to the therapist when he accidently met her at the hospital canteen.

This case illustration throws light on two ethical issues the therapist had to handle. Firstly, was it ethical to involve a third person to motivate the estranged wife to attend therapy and doing so, did the therapist lose her position of neutrality? The second ethical question concerns whether the therapist should accept Mr. Q's gesture of gratitude. It is important to remember the context in which this interaction occurred. Mr. Q belonged to an economically deprived background and his gesture was an effort to express his gratitude to a person who he felt had empowered him. His motivation was not to manipulate therapy but a token gesture that his culture expects of him. In this situation, the therapist defined the boundaries and gently declined the offer.

Although the approach is typically cautious, some therapists do accept gifts on the assertion that it would help promote client's self-acceptance (Knox 2008). Herlihy and Corey (1997) outline circumstances and considerations for accepting gifts; these include the value of the gift, as well as the culture and context in which it is offered.

6.8 Multicultural Competence in Couple and Family Therapy

India is marked by remarkable cultural diversity; there are variations across regions, religions and socioeconomic strata. A point to note here is that most of the trained therapists are from an educated, upper middle class with an urban background.

Training in family therapy is predominantly modelled after the Western literature and values. At the same time, the values of family therapists in India are embedded in the larger culture to which they belong (Nath and Craig 1999).

The lack of competence in dealing with cultural differences that crop up in the course of therapy is a major ethical concern. Poor awareness of the culture and of the prevalent familial practices would lead to wrong assessments and misinterpretations of the functioning and dynamics of the family. For instance, arranging a marriage to the maternal uncle or maternal cousin may be acceptable in some communities in South India. Similarly, the matriarchal system is prevalent in many communities in the states of Kerala and Meghalaya. Therapists from urban settings could find it difficult to understand and address the power structures, interactions, communications and gender roles in a rural and more traditional setting. We would like to bring into focus one of the author's experiences with a new cultural setting to highlight the issue of multicultural competence.

The therapist took up a job at a small town in the northeast region of India. Having lived in a big, cosmopolitan city for most of her life, she was never exposed to families with tribal background. In the initial days of work, she had to work with a family from one of the northeastern states. The wife in this family was the person receiving psychiatric treatment. Her husband was married seven times and one of the wives was the client's own younger sister. The husband was taking turns to live with each of these women from whom he had many children. Most of the wives were living in close proximity to each other, each working to feed their children, sometimes involved in each other's care, while the husband was mostly unemployed. She found it difficult to understand this arrangement and to find a way to work with this family system. She sought the help of peers from the region to understand and work with the complexities of this family system.

Therapists working with those culturally different from themselves need to be aware of their own responses and reactions to the difference, and be cognizant of their incompetency in dealing with differences. It is the ethical duty of the therapists to update themselves with the skill and knowledge to widen the horizons of their competence.

Indian families are going through several changes. The composition of families in India is constantly being redefined and continues to be diverse in their structure, relationships and language and with myriad influences from larger social issues and problems. Therapists working in India have to be competent in understanding rural and urban differences, balancing between 'traditional' versus 'modern' attitudes in a given family, learn to literally speak the language of the family and be adept at handling the variations in family interactions across India (Shah et al. 2000).

The applicability of family therapy models and interventions developed based on Western understanding is a debatable issue. Viewing the Indian family through the Western 'eye' and understanding could be counterproductive (George 2013). Sciarra and Simon (2008) consider the imposition of an external conceptual framework on the notion of family as a form of colonialism. They propose a mix of universal and traditional models in order to balance the universal aspects and the culturally unique issues of a family. Family therapy models must reflect the cultural context and relevant models and techniques may need to be developed.

6.9 Ethics of Therapy and Legal Directives: A Conundrum

While professional ethical codes and guidelines and codes steer family and couple therapists in the negotiation of complex ethical issues, the intersections with legal frameworks can introduce newer challenges.

The issue of maintaining confidentiality in the context of one partner disclosing HIV status during the course of therapy is a complex one. Even though professional ethical guidelines clearly respect the individual's right for privacy and confidentiality, the law of the land needs to be put above these guidelines. A Supreme Court of India ruling in the case of HIV/AIDS clearly states that the right to privacy and confidentiality is not absolute (Mathiharan 2002). The rights of the third party also need to be considered and protected. The National AIDS Control Organisation (2008) also recommends that persons with HIV be encouraged to notify their partner/s.

The proposed HIV/AIDS Bill (2014) of India lays the following conditions to physicians and counselors for partner notification:

(a) When there is reasonable indication that the partner is at significant risk of transmission of HIV from such person; and
(b) The person with HIV has been counselled to inform his/her partner
(c) If the healthcare provider has reason to believe that the person with HIV will not inform his/her partner
(d) The healthcare provider has informed the person with HIV his/her intention to disclose the HIV-positive status to such partner; and
(e) Such disclosure to the partner is made in person and with appropriate counselling or referrals for counselling.

The Bill also states that the healthcare provider is not obligated to locate the partner. It further states that the decision to disclose or not disclose should not result in any litigation against the healthcare provider. In case of women and children, the healthcare worker can decide not to disclose if there is reasonable ground to suspect that this information would result in violence or abandonment and thus be a threat to their safety.

The Protection of Children against Sexual Offences Act, 2012, of the Government of India has a provision that makes reporting of anticipated and actual child abuse mandatory by medical professionals and others. A failure to report would result in punishment with a fine, with imprisonment up to six months or both (Gupta et al. 2013).

Mandatory reporting puts therapists in a situation where they have to breach confidentiality. The likelihood of clients feeling cheated, losing trust and dropping out of therapy is very high. The issues that were dealt in therapy now have to be reported to the police and this could clearly disrupt therapy. The welfare of the family, especially that of child/person who is affected by this decision, is a major concern for the therapist. The supporting system in India that deals with the family or child after mandatory reporting has inadequate provisions to handle effects that this could have on the child and the family. The adverse events that could ensue in the family after mandatory reporting are not discussed anywhere in the law and there is no clear direction about how to deal with the rest of the family system once such a report has been made.

Family and couple therapists in India face a unique situation while dealing with issues of divorce or separation in therapy. Most religions view marriage as a sacrament and divorce is seen as a stigmatizing and objectionable affair (Rao and Sekar 2002). Though the divorce rate in India is very low when compared to the West, it is on the rise (Nambi 2005). Family and couple therapists are often confounded with the question of addressing the issue of divorce within the therapy context. The therapist should also be aware of his/her own values and thoughts on marriage, separation and divorce (Margolin 1982). The therapist's role is to act as a guide who helps the family decide what questions they want to ask, which particular behaviours they wish to change, where they can go to get the answers and relevant skills, and encourage them to do what they feel must be done (Broder 1975).

Family laws in India are influenced by religious traditions and inadequate awareness of the various laws can pose serious problems in the conduct of therapy. For example, Muslims in India are governed by their personal laws which permit a man to seek divorce under a range of circumstances, while the conditions laid down for women are quite different. Mediating succession and property issues in interfaith marriages also brings into question the dilemma of personal law versus the Special Marriage Act of 1954. For example, a couple of interfaith marriage planned to divorce. The wife had converted to her husband's faith after a few years. The wife wanted an amicable settlement. Despite being married under the Special Marriage Act 1954, the in-laws insisted that the wife follow the traditional laws of her husband and refused to support her in any way. Ignorance of the law and insistence on following traditional laws was a challenge for mediating therapist.

An ethical choice is called for when it comes to self-determination of the individual versus the family's collective decision or stand on divorce or separation. This is often seen when one spouse wants to exercise his or right for divorce while the other spouse and family remain totally against this. There are occasions when one spouse looks for a mutual divorce while the other wants to contest the divorce in court. Here, the cultural and professional competence of the therapist is paramount. The therapist is faced with a further dilemma when there is no competent agency to refer the couple for mediation or decision-making, or when existing structures make these services inaccessible due to exorbitant consultation fees.

Working within these legal frameworks, which are sometimes at odds with ethical guidelines or with therapists' personal values, may place the practitioner in an ethical conundrum. Ethical values of confidentiality, responsibility and informed consent may conflict with the extant laws. Couples and families are likely to stop placing their trust in therapy once there is a breach in confidentiality. A prior discussion on the likelihood of breach in confidentiality and informed consent could perhaps help some families understand this requirement.

6.10 Conclusion

It is increasingly clear that in the absence of a clear ethical guideline from any competent authority, Indian couple and family therapists have to heavily depend on those from the West. Most often these do not offer solutions to conditions that are specific to this culture and the socio economic realities. The ethical issues of maintaining boundaries and confidentiality become more prominent and challenging in the Indian diaspora, which is primarily an interdependent culture. Therapists have to evolve their own mechanisms of arriving at ethical decisions. Being an ethically sensitive therapist requires one to don many hats within the therapeutic context. The therapist may often have to choose to remain silent on issues on which he/she may have a strong opinion, and at the same time, have to become the voice of the most vulnerable member of the couple/family. Making ethically sound decisions requires the couple/family therapist to remain alert to the dynamics and being reflective about one's position. A special care has to be taken to discuss ethical issues in the conduct of family therapy during supervision. The supervisory sessions should accommodate discussions on ethical dilemmas and train prospective therapists in ethical decision making (Bhola et al. 2015). Family therapy training in India should be standardized at the same time incorporate region specific issues and competence. It is clear that ethical aspects in couple and family therapy need forums for dialogue and debate.

References

Ables, B. S., & Brandsma, J. M. (1977). *Therapy for couples*. San Francisco: Jossey-Bass Inc.

Agrawal, J., & Shah, A. (2009). Couples therapy and extra-marital involvement: Principles and practices. *Indian Journal of Social Psychiatry, 25*(3–4), 78–88.

Almeida, R. (1996). Hindu, Christian and Muslim families. In M. McGoldrick, J. Giordano, & J. Pearce (Eds.), *Ethnicity and family therapy* (2nd ed., pp. 395–424). New York: Guilford Press.

American Association of Marriage and Family Therapy. (2012). *Code of Ethics*. Retrieved April 16, 2016 from https://www.aamft.org/imis15/Documents/AAMFT%20Code_11_2012_Secured.pdf

American Psychological Association. (2010). *Ethical principles of psychologists and code of conduct*. Washington, DC: Author. Retrieved June 8, 2015 from http://www.apa.org/ethics/code/index.aspx

Avasthi, A. & Grover, S. (2009). Ethical and legal issues in psychotherapy. *Indian Journal of Psychiatry: IPS Clinical Practice Guidelines*. Retrieved January 5, 2016 from http://www.indianjpsychiatry.org/cpg/cpg2009/article11.pdf

Bennett, B. E., Bricklin, P. M., Harris, E., Knapp, S., Vandecreek, L., & Younggren, J. N. (2007). *Principles of risk management: A patient-oriented approach*. Washington, DC: APAIT.

Bhola, P., Sinha, A., Sonkar, S., & Raguram, A. (2015). Ethical dilemmas experienced by clinical psychology trainee therapists. *Indian Journal of Medical Ethics, 12*(4), 206–212.

Broder, E. A. (1975). Assessment: the foundation of family therapy. *Canadian Family Physician, 21*(10), 53–55.

Cannon, C., & Buttell, F. (2015). Illusion of inclusion: The failure of the gender paradigm to account for intimate partner violence in LGBT relationships. *Partner Abuse, 6*(1), 65–77.

Corey, G., Corey, M., & Callanan, P. (2011). *Issues and ethics in the helping professions*. CA: Brooks and Cole/Cengage Learning.

Duggal, C. (2014). Negotiating an open marriage in couple therapy. *Psychological Studies, 59*(1), 76–81.

Gambrill, E. (2008). Evidence-based (informed) macro practice: Process & philosophy. *Journal of Evidence-based Social Work, 5*(3–4), 423–452.

George, W. A. (2013). A paradigm shift in family therapy in India: Exploration from socioeconomic, cultural and spiritual perspectives. *International Journal of Physical and Social Sciences, 3*(3), 153–166.

Goldberg, H., & Goldberg, I. (2008). *Professional issues and ethical practices—family therapy: An overview* (7th ed.). Canada: Thomas Books/Cole.

Grove, J., & Blasby, S. (2009). The therapeutic encounter in same-sex couple counseling—the client's perspective. *Counseling and Psychotherapy Research, 9*(4), 257–265.

Gupta, N., Aggrawal, N. K., & Bhatia, M. S. (2013). The protection of children from sexual offences (POCSO). *Delhi Psychiatry Journal, 16*(2), 429–431.

Gurman, A. S., & Kniskern, D. P. (1978). Research on marital and family therapy: Progress, perspective, and prospect. In S. L. Garfield & A. E. Bergin (Eds.), *Handbook of psychotherapy and behavior change: An empirical analysis* (2nd ed., pp. 817–901). New York: Wiley.

Herlihy, B., & Corey, G. (1997). *Boundary issues in counseling: Multiple roles and responsibilities*. Alexandria, VA: ACA Press.

Indian Association of Clinical Psychologists. (2015). *Ethics and code of conduct of clinical psychologists Guidelines 2012–13*. Retrieved June 19, 2015 from http://www.iacp.in/node/159

Jackson, A. P., Hansen, J., & Cook-Ly, J. M. (2013). Value conflicts in psychotherapy. *Issues in Religion and Psychotherapy, 35*(1), 6–15.

Knox, S. (2008). Gifts in psychotherapy: Practice review and recommendations. *Psychotherapy: Theory, Research, Practice Training, 45*(1), 103.

Mathiharan, K. (2002). HIV and AIDS: some legal and ethical implications for the medical profession. *Indian Journal of Medical Ethics, X, 4*, 79–82.

Margolin, G. (1982). Ethical and legal considerations in marital and family therapy. *American Psychologist, 37*(7), 788–801.

Ministry of Health and Family Welfare. (2014). *The human immunodeficiency virus and acquired immune deficiency syndrome (prevention and control) bill, 2014*. Retrieved May 13, 2016 from http://www.prsindia.org/uploads/media/HIV/HIV-AIDS%20Bill,%202014.pdf

National AIDS Control Organisation. (2008). *Ethical guidelines for operational research on HIV/AIDS*. Ministry of Health and Family Welfare, GOI. Retrieved May 13, 2016 from http://www.naco.gov.in/NACO/About_NACO/Polivu_Guidelines/Policies_Guidelines1/

Nath, R., & Craig, J. (1999). Practising family therapy in India: how many people are there in a marital subsystem? *Journal of Family Therapy, 21*(4), 390–406.

Nambi, S. (2005). Marriage, mental health and the Indian legislation. *Indian Journal of Psychiatry, 47*(1), 3–14.

Pachankis, J. E., & Goldfried, M. R. (2013). Clinical issues in working with lesbian, gay, and bisexual clients. *Psychology of Sexual Orientation and Gender Diversity, 1*(S), 45–58.

Rao, A. B. S. V. R., & Sekar, K. (2002). Divorce: Process and correlates, a cross cultural study. *Journal of Comparative Family Studies, 33*(4), 81–104.

Sciarra, D. T., & Simon, G. M. (2008). Assessment of diverse family systems. In L. A. Suzuki & J. G. Ponterotto (Eds.), *Handbook of multicultural assessment: Clinical, psychological, and educational applications* (3rd ed., pp. 247–272). San Francisco: Jossey-Bass.

Shah, A., & Satyanarayana, V. A. (2011). The solo note: Therapy with one spouse through systemic individual marital therapy (SIMT). *Journal of Family Psychotherapy, 22*(4), 297–312 http://dx.doi.org/10.1080/08975353.2011.627800

Shah, A., Varghese, M., Udaya Kumar, G. S., Bhatti, R. S., Raguram, A., Sobhana, H., & Srilatha, J. (2000). Brief family therapy training in India: A preliminary evaluation. *Journal of Family Psychotherapy, 11*(3), 41–53.

Strom-Gottfried, K. J. (2008). Values and ethics for professional social work practice. In K. M. Sowers & C. N. Dulmus (Eds.), *Comprehensive handbook of social work and social welfare* (pp. 101–121). New York: John Wiley and Sons.

Tamura, T., & Lau, A. (1992). Connectedness versus separateness: applicability of family therapy to Japanese families. *Family Process, 31*(4), 319–340.

Thomlison, B., & Corcoran, K. (2008). *Evidence-based internships: A field manual*. New York: Oxford Press.

The Parliament of India. (1954). *The special marriage act*. Retrieved on May 13, 2016 from http://keralaregistration.gov.in/pearlpublic/downloads/The%20Special%20Marriage%20Act.pdf?tok=49sddh3ss34ff4

Wells, R. A., & Giannetti, V. J. (1986). Individual marital therapy: A critical reappraisal. *Family Process, 25*, 43–51.

Younggren, J. N., & Harris, E. A. (2008). Can you keep a secret? Confidentiality in psychotherapy. *Journal of Clinical Psychology, 64*(5), 589–600.

Chapter 7
Ethical Issues in Working with Women in Sex Work Facing Intimate Partner Violence

Sanghamitra Iyengar

Abstract This chapter explores the ethical issues facing counsellors in working with special populations. The focus is on women in sex work in the community setting in India who face intimate partner violence (IPV). It describes the complex nature of intimate partner relationships in the lives of these women, the violence within these relationships and the resistance to disclose or report it. It identifies the hidden nature of the violence and the reluctance of those experiencing it, to either step out of it or take action against the perpetrators as one of the key challenges for counsellors and therapists working with this population. The chapter illustrates how despite every episode of violence being a crisis, the conventional crisis response itself can pose a threat to the safety of the client. The author traces how the context of IPV in this population differs from that faced by other women and proposes a seven-step, multisectoral model of practice that ensures autonomy, non-maleficence, beneficence and justice.

Keywords Women in sex work · Intimate partner violence · Resistance to report · Ethical issues · Counselling · Multisectoral model of practice

Intimate partner violence (IPV) is a complex phenomenon. Embedded in patriarchy, gender role and power differentials, it is made complex by its roots in intimacy, privacy, social identity and social position. All these hamper its visibility. Affected persons strive to keep it hidden. Protection and rescue are difficult, let alone redressal and justice.

These very factors lie at the root of most of the ethical challenges faced by counsellors working with IPV. While acknowledging the fact that there has been a gender bias in research on IPV, this chapter will limit itself to exploring the

---

S. Iyengar (✉)
Samraksha, Bengaluru, India
e-mail: sanghamitra.iyengar@gmail.com
URL: http://www.samraksha.org

P. Bhola and A. Raguram (eds.), *Ethical Issues in Counselling and Psychotherapy Practice*, DOI 10.1007/978-981-10-1808-4_7

violence on women by their male partners. It will, specifically, look at women in sex work and their unique situation in IPV as being different from that of other married women in many ways. The author recognizes that transgenders and other sexual minorities face similar violence from partners assuming the male gender role, but the scope of this chapter will be limited to the ethical issues faced by counsellors working with women in sex work in the context of IPV. This chapter has used case scenarios and self disclosures to aid in understanding the complexities of many of the issues discussed. It should be noted that all names have been changed to maintain anonymity and to ensure confidentiality of clients involved.

7.1 Intimate Partners in the Life of Women in Sex Work

It is estimated that there are 868,000 sex workers in India of whom 734,000 are reached by programmes related to HIV prevention (NACO 2013), where they have some access to counselling and crisis intervention. This is a heterogeneous population with diverse profiles, spread across the country. However, several studies report that irrespective of setting, a high proportion of sex workers have a regular, non-commercial, intimate partner (Bhattacharjee 2013; Deering et al. 2011).

It is reported that in brothel-based settings, clients wanting exclusive service turn into live-in partners/spouses. Gradually, however, they succumb to the lure of the sex worker's earnings and become agents or pimps focusing largely on what the sex worker can bring in (Somanath et al. 2013).

This chapter will look at the intimate partner relationships of women practising sex work in the community setting. In this context, two types of partners are seen: husbands (legally married) or live-in or regular, long-term sexual partners who are presented as husbands to the social environment. Violence has been reported from both these settings (Karandikar 2008).

Legally married partners could be those to whom the women were already married before they started sex work. They report moving to sex work when their husbands could not meet their need for love, sex or economic necessities. The following disclosure illustrates the complex challenges such a relationship evokes in the lives of sex workers.

> My husband was a strange man, not interested in anything. He would not go to work and be just drinking. All my jewellery was pawned, the little land that we had was sold, and I had to go and take credit from the nearby provision store. One day, the store owner refused to give any more credit and when I begged him, he asked me to sleep with him. Gradually, I found more people who were ready to pay for sex, and earned enough to run the house. My husband never asked me where the money was coming from, as long as I left him alone. He never stops me from going out, but three four times a month, he shows his manhood and beats me up. I just keep quiet. Both of us know that I need to do sex work to keep the house running, but he has to keep his prestige as a man.

Sometimes, the men are reportedly unaware of their wives' sex work practice and the violence that occurs is stated to be about 'not giving enough time to children', 'not taking care of the house', and 'dressing up and going out too much'. At other times, the men know but do not question it directly, as they are dependent on the women economically. Often, the women report silence on the sex work issue, but violence on other pretexts such as those stated above. They also report coercive and violent sex.

Legally married partners could also be those whom the women meet as clients, form a mutual attachment and get married. Here, the ongoing sex work may continue to be hidden or the client could take control of the sex worker's earnings (Karandikar 2008). Regular clients could also take on the role of a 'spouse'. They do not usually pay directly for sexual services but support the sex worker in other ways. Often, they pay the rent or school fees of the children and pay regular maintenance money (Orchard 2007). Many of them become live-in partners.

7.2 Relationships of Women in Sex Work with Their Clients Versus Intimate Partners

While women in sex work have clients to whom they offer sexual services, they consider it a professional relationship in which they draw boundaries (Maher et al. 2013) and do not get into an emotional relationship with them. While they may face violence from clients, they largely avoid clients that are violent and have several strategies to deal with that (Sanders et al. 2009). Today, with better awareness of the law and the support of collectives, they are much more assertive and prepared to act on the violence from clients.

Except in some coercive brothel situations, sex workers actually challenge patriarchy. In the commercial relationship with their clients, they have a fair degree of control. They have sex outside marriage; choose whom they offer their service to; set rates for different services, and decide when and what type of sex they will provide. There is risk of violence and coercion, but by and large, community-based sex workers have better control in the relationship with their client.

With their non-paying, regular partners, however, the situation completely changes. The relationship turns intimate and the partners take on the role of a spouse. Both slip into a traditional marital dyad kind of relationship and the power equation changes. The intimate partner often lives in the house, plays the father to the children and may provide for the household. There is some attachment to begin with, but women largely report that it gets converted to control: emotional, social, psychological and economic control and often ends in violence. It is necessary for counsellors to understand the underlying needs of the women which keep them in these relationships despite the violence.

7.3 The Need for Love

Sex workers in their business, need to move away from the traditional concept of love and intimacy accompanying sexual interactions. Narratives of sex workers reveal that they are able to offer sex for payment without love and intimacy (Maher et al. 2013). However, they continue to aspire for romantic love in a relationship (Warr and Pyett 1999). As has been pointed out (Aveling et al. 2009), love offers "a positive identity, a sense of being valued and a promise of power from being loved". In the world of stigma and social rejection that women in sex work live in, they are not valued or considered 'worthy' even though they are paid for their services. It is 'love' that provides a validation of the self. The inner need to attain and retain this love, leads women in sex work into vulnerable positions in these relationships. They report that intimate partners take emotional and economic advantage of this vulnerability. When the partner begins to exercise more and more control over the behaviour of the sex worker and her financial assets, she resists and it results in violence (see Fig. 7.1).

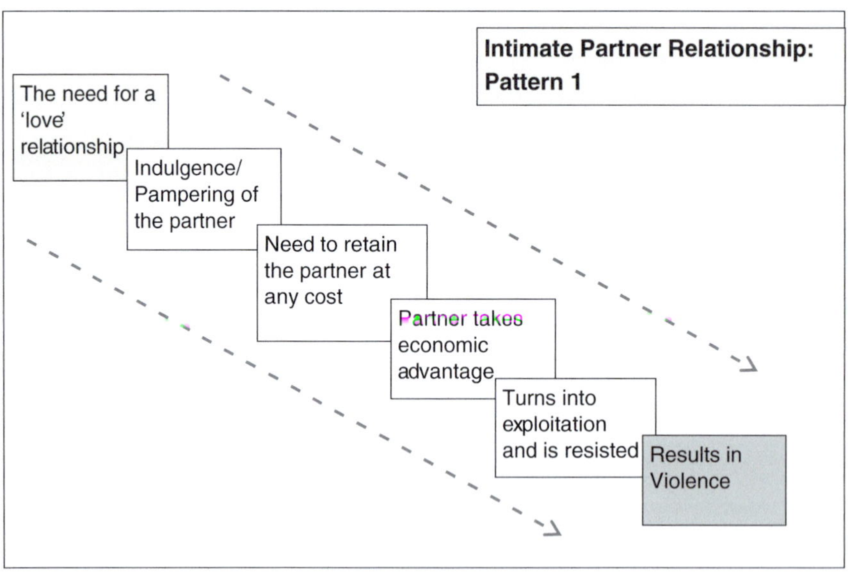

Fig. 7.1 The association between the need for love and violence

7.4 The Need for a Social Identity

The sex worker identity is a highly stigmatized one in India as in many parts of the world. The last two decades of empowerment initiatives through HIV prevention programmes have helped women in sex work to perceive sex work as work. In a rights-based environment or in the context of their collective, they have been able to claim their identity with pride. In their identity as sex workers, they challenge the patriarchal positions on women's sexuality and sexual behaviour. However, in the larger society, they desire other identities that are non-stigmatized and can give them social power and social status. This need to carve out other identities which have social sanction is very strong. One such identity is that of a 'family' woman: a wife and a mother.

This leads them to create family-like structures with their intimate partners, who pose as their spouses even when they are not married. This gives them the 'normalcy' in their social environment, and the possibility of social participation and recognition. At another level, it helps their children to have a father figure. Whether the intimate partner in his role of the father is nurturing or not, he is at least physically present. This prevents social ostracism and gives them and their children a social space and social acceptance.

This, however, comes at a price. The intimate partners often take a patriarchal position and exercise their gender power as 'spouses' demanding special privileges. Women in sex work, who live in the community, are dependent on the partner to maintain the 'family' image. The partners take advantage of this vulnerability and there are numerous narratives of blackmail, abuse and violence (see Fig. 7.2).

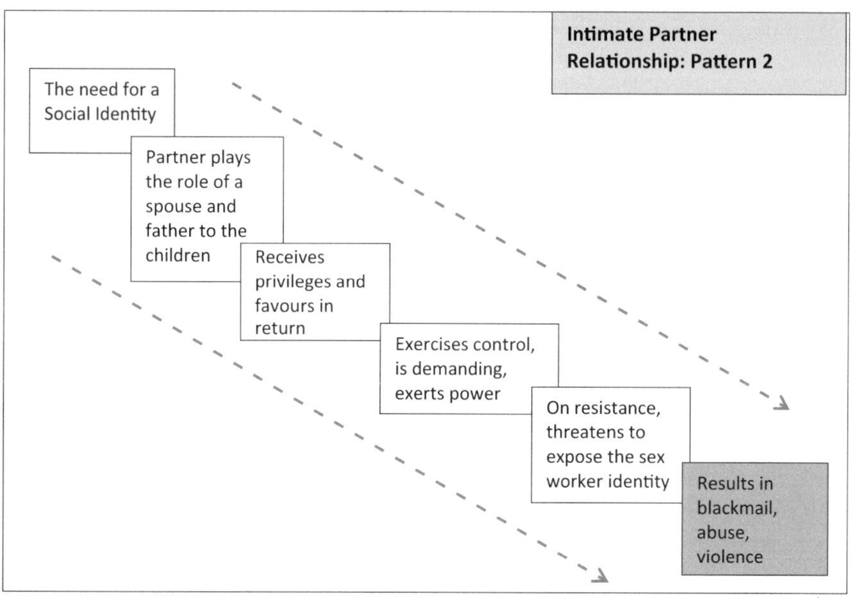

Fig. 7.2 The association between the need for a social identity and violence

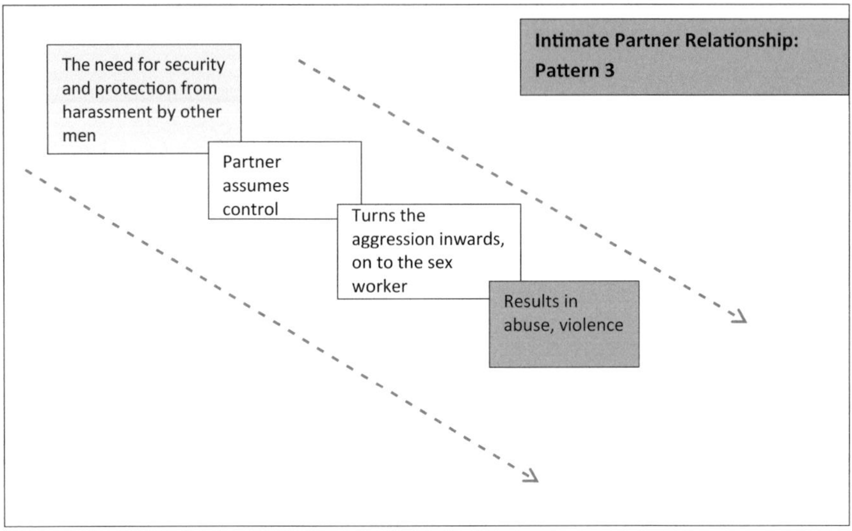

Fig. 7.3 The association between the need for security and violence

7.5 The Need for Security

Intimate partner relationships also meet a need for security from other threats in society. Sex workers report that they need a 'husband' figure to control intrusion and coercion from unwanted clients as well as other men in the neighbourhood. They share that they often get drawn into partner relationships with a 'tough' figure to keep others at bay, but this sometimes leads to losing control over their personal lives. The partners begin to exercise control in all spheres of their lives—domestic, economic and sexual. Non-compliance is punished with violence. Freedom is curtailed and behaviour is monitored. The aggressive stance needed for security, thus, often turns into aggression towards the sex worker. Demands for autonomy lead to further violence (Reza-Paul et al. 2012) (see Fig. 7.3).

7.6 Resistance to Report Intimate Partner Violence

Evidence from a large cross-cultural survey in India reveals that around 54 % of female sex workers did not disclose their experience of violence to anyone, with considerable variations in the pattern of disclosure across states. Another 36 % of them shared the experience with NGO workers/peers (Mahapatra et al. 2014). Compared to the violence perpetrated by paying partners/strangers, that by a non-paying partner was twice more likely not to be disclosed (Mahapatra et al. 2014).

Fear is a major factor preventing women in sex work from reporting violence. However, in India, over the last two decades, collectivization processes have instilled courage in the women to report the violence perpetrated by police, *local thugs* and even clients. However, when it comes to IPV, there is a strong resistance to report, even among the more empowered of the sex workers.

Exploring what women in sex work seek from intimate partner relationships helps us to understand this resistance, and the fear of loss underlying it better, this loss spans psychological, emotional, economic and social realms. To some extent, these overlap, but it helps to examine each realm separately.

7.6.1 Threat of Loss: Psychological

Most women in sex work have experienced several traumatic experiences. They have had to pass through multiple rejections, leading to a poor sense of self and self-worth. The nature of their stigmatized profession itself leads to a negative image of themselves as reflected back to them by clients, friends and family. To offset 'the social transgression' of having multiple partners, they choose one 'special' or intimate partner with whom the relationship is different; a relationship in which they enact the traditional wife's role, often submissive and dependent, where sex is not sold, but given freely. This provides personal validation and a more acceptable sense of self. The violence from partners is interpreted as having an underlying script of 'you are mine' which provides a sense of belonging and feeling valued and wanted. Reporting the violence could lead to the loss of the partner. Losing the partner, might mean losing this positive sense of self.

7.6.2 Threat of Loss: Emotional

Despite differing opinions, literature points to sex, love and intimacy being closely connected in women (Peplau 2003; Warr and Pyett 1999). In the profession of sex work, sexual service is a transaction. The agreement in each transaction is to provide sexual service in exchange for money, but there is no clause for the provision of love. As mentioned earlier, despite engaging in sex with many men, women in sex work draw their boundaries and try not to get emotionally involved with their clients (Maher et al. 2013). This protective measure to prevent complications that could leave them lonely and unhappy. They seek romantic love in an exclusive relationship. Having a regular or live-in partner meets some of these emotional needs. Having this object of love appears to be as important as being loved. Fear of losing this person leads to tolerance of the violence and silence on it.

7.6.3 Threat of Loss: Social

As mentioned earlier, aspiring to grow as close to the socially sanctioned role of 'a married woman' is a psychological and social strategy against stigma. Having a live-in or regular partner can help to mask the sex work identity and create an impression of normalcy to people in her social network. Such an arrangement also provides some benefits to the children, who can claim a father and a conventional family unit. It also offers access to the social status, privileges and opportunities for social interaction as well as community resources available to married women/family women, but denied to sex workers. These could include community leadership, ownership of land, participation in social and religious events, and use of community resources, schooling and other opportunities in the mainstream for their children. Reporting the violence puts at risk the secrecy of the sex worker identity. What is at stake is social status and prestige. On another dimension, there is fear that any action against the partner could result in danger to their children or their being taken away from them.

7.6.4 Threat of Loss: Economic

Although it is true that women in sex work often support their live-in partners, economically, it is also true that in a large number of cases, the partners provide economic support that stabilizes the daily living. Sex work income is not steady and has ups and downs. Savings and reserves by sex workers are largely absent (Mahila Kranti 2013) and reporting the violence could lead to a loss in the relationship and economic provision by the partner. Additionally, the reporting could lead to the disclosure of identity which could also make the practice of sex work difficult (for example, reduction in partners, as hidden clients do not like to visit 'open' sex workers; or social sanctions; or a close watch by the law).

7.7 The Complexity of IPV Faced by Women in Sex Work

Although reluctance to report IPV is seen both in married women and sex workers, in the former, the relationship is visible and publicly claimed. To a large extent, there is some degree of family engagement and support. With women in sex work, on the other hand, the relationship is more fragile as it is not legally or socially binding most of the time. Reporting the violence may lead to abandonment and loss of support. Additionally, the sex worker may have to face stigma with little hope of redressal.

The precarious social situations, the secrecy in which the relationship is shrouded, the stigmatized identity, all contribute to a high level of vulnerability for women in sex work. Families and other social connections are kept at bay to maintain the myth of a conventional married relationship. This cuts away the social supports and protective mechanisms against violence. Physical abuse and violence are common and their visibility sometimes draws some support, even where the women do not report them. But, in sexual and psychological violence, women in sex work are left with very few options for support. The following case scenarios (see Box 7.1) to reveal the complexities discussed above.

Box 7.1: Case Scenarios 1

Kanaka is a 36-year-old sex worker, who has been living with her partner for the past 10 years. He has been economically demanding and controlling and she has become more assertive in the last few years. Kanaka has an 18-year-old daughter for whom she recently fixed a marriage alliance. The daughter kept opposing the marriage, but was persuaded by the 'stepfather' to get married. Six months after the marriage, the daughter returned home and disclosed to the mother that the stepfather had been having a relationship with her since she was 15. She loved him. She had not wanted to get married, but he had told her that it would provide a better cover for their relationship if she got married. But, she had returned as she wanted to be full time with him. She told her mother that she could satisfy his needs better than her. When Kanaka confronted him, he spoke of paying her back as she had not been 'co-operative'. He has threatened her with harm to her daughter and public disclosure if she tries to take away the daughter. Kanaka is psychologically shattered. She is full of guilt at what has happened to her daughter, emotionally devastated at having lost her partner and extremely worried about what is in store for her daughter. She says she has suddenly lost all that mattered to her in one stroke: both her partner and her only daughter.

Saina is a 25-year-old sex worker who married one of her clients in a temple. Initially, she stopped sex work and became a housewife. After 6 months, he started bringing clients to her, many of whom were abusive. He fixed the rates and took the money. When she resisted, he threatened to tell all the neighbours about her. After 2 years of this, she met some women from a sex worker collective and she went to their office to talk about her plight. When he got to know this, he took away all her clothes and locked her up naked, so that she could not leave the house. He sexually abused her as well.

Mary, 29, is in a live-in relationship with one of her former clients. He provides well for her financially, shows love and does not use physical violence. But he prevents her from being in contact with her friends or family and from leaving the house. He has taken away her phone and forbidden her from talking to anyone else. He threatens to leave her if she meets anyone else. She is emotionally dependent on him and cannot think of breaking away, but feels that she is losing her confidence and capacity for everything.

Kala is 28 years old with a 6-year-old son. Kala's partner is alcohol dependent and of late has become very violent towards her. He has threatened that he will reveal her sex work identity to the school authorities where her son is studying if she tells anyone about the violence.

The instances in Box 7.1 reflect the complex situations, and the difficulty in giving up what exists despite the violence.

7.8 Ethical Challenges in Counselling Women in Sex Work Facing IPV

The resistance to talk about the violence and seek help is obviously the very first challenge that the counsellor is faced with. How does one bring the person facing violence into a help-seeking situation, if the person does not want to talk about it? The dilemma for the counsellor is the conflict between respecting client self-determination and client protection.

In situations where the sex worker decides to approach the counsellor, or has disclosed to a confidant who has brought her to the service, or has been referred from a hospital/shelter setting where she has reached, there are challenges in taking the history. What can be the psychological impact of the recall of trauma? What can be the social impact of disclosure? Should the counsellor try and facilitate disclosure? Or should she/he respect the desire of the client not to disclose completely or even accept versions exonerating the perpetrator? How does the counsellor handle the same dilemma of respecting autonomy versus ensuring safety?

If the client gives the history, but wants to maintain status quo in the relationship, how should the counsellor respond? The status quo might mean stepping back into the violence. How can the counsellor respond ethically when respecting the client's wishes might mean sending her back into a harmful situation or even a life threatening one?

Where the client is prepared to act, how does one ensure that the path to action to be pursued by the client (complaint registered at the police station, arrest of the partner etc.) does not have other adverse effects on the client's life?

In short, how can the counsellor ensure all the ethical principles of autonomy, non-maleficence, beneficence and justice? Box 7.2 is depictive of the ethical dilemmas posed while counselling women in sex work facing IPV.

Box 7.2: Case Scenario 2

Gouri was a 31-year-old woman in sex work, living with her 10-year-old son and partner. She worked in an HIV prevention programme advising other women in sex work to practise safe sex and take care of their health. Her partner was educated and worked in the same town. Over the last year, he developed suspicions regarding Gouri travelling to other villages to meet the women, and although he did not discuss it in the day, every night he accused her of infidelity and beat her severely.

At the workplace (the sex worker collective), her injuries were noticed and she was probed about it, but she was not ready to talk about the violence. One day, she was beaten badly and her arm was fractured. She, then, opened up to the counsellor. She just wanted to stop working and stay home. She felt that the work situation was creating the conflict.

She stopped working for a few months, reported reduction in the violence and came back. She said her partner was agreeable to her working again. But once she did the violence recurred. Her peers offered support to challenge the partner and issue him a warning. Gouri expressed fear that he might leave her and she would not find another partner. The violence was only at night. She could send her child out of the house and manage. As the violence increased, after discussion with her, the senior leaders of the collective met with her partner and issued him a warning that they would lodge a First Information Report (FIR) at the police station if the violence recurred. He expressed regret to the group and assured them that he would take good care of Gouri. Following this, Gouri reported that the violence had abated. A few days later there was another incident, but Gouri refused consent to report it or for anyone to come home again. The crisis team and the counsellor tried to persuade her, but she was extremely fearful of the repercussions and refused to take any action. The leaders of the collective posted her to another town, hoping she could be far away from the abuse.

Two days later, there was news that Gouri was in the hospital with third degree burns. She died that night. The investigation is on as to whether it was suicide or murder.

What went wrong? Should the counsellor and crisis team not have intervened? Should they have intervened earlier? Was sending her away from the partner, which was also away from her support network, the action that led to this?

There are no easy answers, but the case story, highlights the fragile and extremely volatile context of IPV. There is enough evidence that violence drains away the agency of the women experiencing it, leaves them feeling fearful, helpless, hopeless and unable to act. Working with the individual, giving her a better sense of self-worth and the confidence to deal with it and providing back up support has been seen to work better than addressing the violence directly.

7.9 The Helping Process in Intimate Partner Violence

From our understanding of the IPV faced by women in sex work in India, it is clear that multiple strategies and types of resources are needed to respond to this complex issue. Broadly, the strategies include: unobtrusive first contact, assurance of safe communication, risk analysis, safety plans, action against violence, support systems to handle repercussions and long-term support to heal.

7.9.1 Step 1: Unobtrusive First Contact

Objective: Opening a Support Channel

The first step is the discreet identification of the violence situation (see Box 7.3). Since women facing violence do not report it, it needs to be picked up and primary disclosure enabled by someone with access to them. Often neighbours, close friends, siblings or parents play this role. For women in sex work, these supports may not always be available. However, today, in India, there are structures created by the HIV prevention programmes which can offer peer support in the community setting. There are community counsellors and crisis teams in almost every HIV prevention programme and there are 508 such programmes in the country to draw on.

Violence depletes personal coping resources and confidence in oneself. The vulnerable situation that women find themselves in, fosters a sense of helplessness, hopelessness and a sense of fatalism. The key intervention here is to reach out and make a helping connection. Confidentiality is paramount. While in crisis interventions, the first priority is moving the client to a place of safety, here it could sometimes be counterproductive, unless the client is prepared for it.

Box 7.3: Key Challenges for the Counsellor: Identifying the Person in an IPV Situation; Facilitating Help Seeking

Intervention Listening, assessing immediate danger, infusing hope, assuring support
Key Counselling Principles Respect, confidentiality, self-determination
Key Ethical Principles Autonomy, non-maleficence

7.9.2 Step 2: Creating Safe Spaces for Communication and Interaction

Objective: Assuring Safety and providing Psychological and Emotional Support

Referrals may come from hospitals or the police when the violence is extreme and has become public; or the sex worker herself may have sought the counsellor out. If there is an open report of violence and identification of the perpetrator, it can move on to the next stage of detailed history taking and risk assessment. If not, the process will need to start from creating a safe environment for disclosure (see Box 7.4).

A safe space to vent and share the trauma and receive emotional and psychological support is critical. This may vary from situation to situation. If the person is unable to leave home, community counsellors will need to provide it at home. When the person is willing and able to access a counselling centre, then this intervention can take place there. Taking safety concerns seriously, reiterating confidentiality and building trust are the critical interventions at this stage.

Box 7.4: Key Challenge for the Counsellor: Building Trust and Communication Channels

Intervention Listening, validation of the individual, communication, providing assurance of confidentiality and ensuring safety.
Key Counselling Principles Confidentiality, self-determination, non-judgemental stance
Key Ethical Principles Autonomy, non-maleficence, beneficence

7.9.3 Step 3: History Taking

Objective: Recording the History of Violence and Analysing Risk

History taking may not be easy or straightforward as the client may travel back and forth in the narration of the traumatic incidents. There may be a lot of suppressed grief emerging, and it may require a few sessions to record the history. However, the counsellor would need to analyse the risks involved, related even to this primary disclosure. Emotional support may be needed throughout this period. Peer support outside the counselling sessions would be a great help.

Despite the possibility of continuing violence, if the client is not willing to take any action, respecting her pace is critical. Often the client's reluctance may be based on intuitive knowledge of repercussions, and till she can plan safety measures, the pace should not be forced. If the client is not prepared for action, she might either retract, or there may be other repercussions. The counsellor will need to deal with the client's anxiety and fear for personal safety and safety of the children. It is probable that the client has lived with the violence for a while before reaching the counsellor and has some temporary strategies to deal with it.

The counsellor has to be mindful of the possible trauma of recall, the humiliation and the low self-worth that the violence may have caused. Asking for unnecessary details of the abuse, particularly sexual abuse would be voyeuristic and could cause further damage to the client's self-esteem.

The analysis of the risks involved in larger decisions such as reporting the violence, filing a case, asking for police protection or leaving the partner has to be discussed jointly with the client (see Box 7.5).

Box 7.5: Key Challenge for the Counsellor: Assessing and Anticipating Internal Trauma and External Dangers in the Context of Reluctance to Report

Intervention Listening, taking history, providing assurance of confidentiality and ensuring safety.
Key Counselling Principles Confidentiality, privacy, non-judgemental stance
Key Ethical Principles Non-maleficence, beneficence

7.9.4 Step 4: Overcoming Ambivalence and Taking Decisions

Objective: Exploring and Strengthening Motivation for Action

The client may have stayed in the violent relationship, for any of the reasons mentioned earlier. It is good to remember that coming out of violence is a process and takes time. Personal preparation is critical and it is important not to rush the person into action (see Box 7.6). Motivation to change status quo, may take time. At this stage, testimonial sharing has proven to be helpful. This can be on a one-to-one basis, or a group sharing with others in a similar situation. Groups are useful, as participants may be at various points in their journey out of the violence and multiple strategies may surface. The ambivalence regarding the violent relationship may also get clarified. Testimonial sharing in the group needs to be followed by support for any possible backlash.

Box 7.6: Key Challenge for the Counsellor: Walking at the Client's Pace; Accepting any Resistance to step out of a Harmful Situation

Intervention Individual and group support through testimonial sharing, offering multiple strategies for action
Key Counselling Principles Respect, non-judgemental stance
Key Ethical Principles Autonomy, beneficence.

7.9.5 Step 5: Drawing up Safety Plans

Objective: Ensuring Safety

Once the client begins contemplating action, safety plans can be drawn up, based on the risk analysis (see Box 7.7). These could span a number of situations: how to get children to a safe place; how to remove personal and precious possessions; finding a safe place to stay on leaving the house; identifying whom to call in an emergency; listing who could be on hand for a physical rescue; how to counter slander and public shaming; and how to deal with social ostracism on disclosure of the sex worker's identity. Often it is also psychological preparation for the loss of the relationship. Mapping social supports jointly with the client, and listing various support resources and shelter options is a good strategy. Crisis teams of sex worker collectives can mobilize a rescue and support group within a couple of hours and drawing on that resource can be considered.

Box 7.7: Key Challenge for the Counsellor: Helping Clients Overcome Resistance to Analyse Risk and Create Concrete Safety Plans

Intervention Practical and psychological preparation for action
Key Counselling Principles Respect, non-judgemental stance, self-determination
Key Ethical Principles Autonomy, non-maleficence, beneficence.

7.9.6 Step 6: Action Against Violence: Accompaniment and Support

Objective: Providing Confidence and Ensuring Safety

A coordinated effort is needed for this stage. A counsellor or social worker needs to help mobilize the appropriate supports. If the client wants to challenge the violence directly, support from peers can be made available. But there may continue to be ambivalence around the decision and the client may retract. Box 7.8 illustrates the ground realities and the need for coordinated effort in taking action against violence.

Box 7.8: Case Scenario 3
Shalini sought the support of the counsellor and some peers to file a complaint with the police and seek police protection against partner violence. The next day she went and withdrew the complaint, saying her partner had assured her that it would not happen again, and she did not want to lose him. This happened two times before she finally took action to move away from him permanently.

If the decision is to ask the partner to leave, the woman in sex work may need back up support, in case there is violence against the client or children. On the other hand, if the sex worker wants to file a complaint at the police station, then peer accompaniment and crisis support are needed. In all cases, physical and practical support may be needed at this point (see Box 7.9).

Box 7.9 Key Challenge for the Counsellor

1. Retraction of the client from the planned action, even at the risk of harm.
2. Lack of resources to protect the client around the clock, from harm resulting through action against the violence

Intervention Physical and psychological accompaniment, coordination with other supports
Key Counselling Principles Respect, non-judgemental stance, self-determination
Key Ethical Principles Autonomy, non-maleficence, beneficence, justice

7.9.7 Step 7: Long-Term Psychological Support

In a review of 75 studies on IPV and its impact, Dillon et al. (2013) found that women who had experienced IPV (physical, sexual, or psychological) consistently had lower mental health and social functioning scores than women who had not. There is enough evidence that systematic and long-term violence destroys confidence, self-esteem and a total sense of self. It leaves deep scars that take a long time to go away. This is an area which has not received enough attention in IPV work. Rescue from the violence is not enough. The healing process is long, and multi-sectoral long-term support is needed (see Box 7.10). The counsellor's role is vital here. Both individual and group therapy may help.

Box 7.10: Key Challenge for the Counsellor: Retention in therapy

Intervention Psychotherapy, emotional support
Key Counselling Principles Respect, non-judgemental stance, self-determination

Through the entire process, professional supervision and support for the counsellor is critical. The autonomy for the client also has to balance her safety and this is a tremendous responsibility and burden for the counsellor. It will be absolutely vital for the counsellor to have regular personal supervision to deal with his or her own emotional and psychological processes.

7.10 The Multisectoral Model

The author offers a multisectoral model (see Fig. 7.4) for practice, especially in the Indian setting. This model has a lot in common with various crisis intervention models as well as models that have evolved to deal with domestic violence.

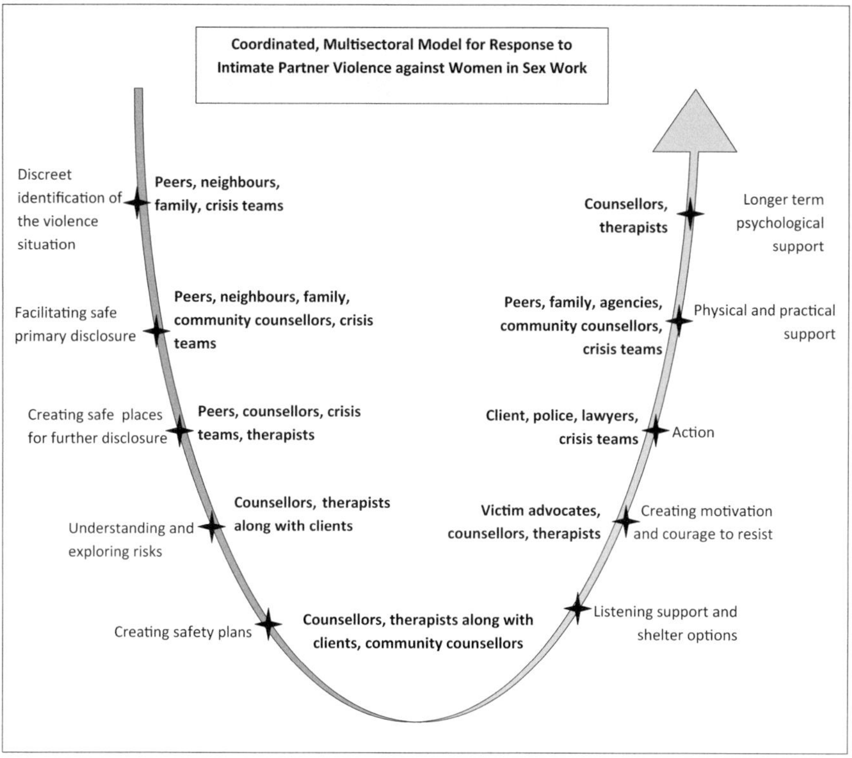

Fig. 7.4 Multisectoral model for response in intimate partner violence against women in sex work

The difference lies in its additional emphasis on autonomy and non-maleficence and inclusion of accompaniment by different players through the entire journey.

7.11 Summary and Conclusion

In the lives of women in sex work, intimate partner relationships take different forms, and serve varied needs. The violence in these relationships is also extremely complex; in its underlying triggers, its manifestation, its isolation from normal societal supports and in the response to it from the victims of violence. What poses key challenges for counsellors is its hidden nature and the reluctance of those experiencing it, to step out of it or take action against the perpetrators. This reluctance is embedded in the gains from the relationship and the socialized tolerance to the violence. Experience in the sector has shown that creating capability to meet the needs currently met by the relationship may reduce the vulnerability and dependence on it. Work with the partners is a direction to be explored.

Every episode of violence is a crisis, but the conventional crisis response poses problems. The journey out of a violent relationship is a long process, and needs the support of a number of people. Therefore, the multisectoral model is recommended.

References

Aveling, E. L., Cornish, F., & Oldmeadow, J. A. (2009). Diversity in sex workers' strategies for the protection of social identity: content, context and contradiction. In B. Wagoner (Ed.), *Symbolic transformation: The mind in movement through culture and society* (pp. 302–333), London: Routledge. Retrieved February 20, 2015 from http://eprints.lse.ac.uk/47797/

Beel, N. (2013). Domestic violence, gender and counselling: Toward a more gender-inclusive understanding. *Psychotherapy in Australia, 19*(4), 44–52. Retrieved February 21, 2015 from https://www.academia.edu/4333682

Bhattacharjee, P. (2013). *Sex work, violence and HIV: Experience from rural Karnataka*. Retrieved September 22, 2014 from https://www.opendemocracy.net/5050/parinita-bhattacharjee/sex-work-violence-and-hiv-experience-from-rural-karnataka

Chacko, S., & Shankaran, L. (2013). Counselling in situations of gender-based violence. In S. Iyengar & D. Sarma (Eds.), *A manual for counselling women in sex work* (pp. 131–164). Mumbai: Tata Institute of Social Sciences.

Chandra, P. S., Satyanarayana, V. A., & Carey, M. P. (2009). Women reporting intimate partner violence in India: Associations with PTSD and depressive symptoms. *Archives of Women's Mental Health, 12*(4), 203–209.

Deering, K. N., Bhattacharjee, P., Bradley, J., Moses, S. S., Shannon, K., Shaw, S. Y., & Alary, M. (2011). Condom use within non-commercial partnerships of female sex workers in southern India. *BMC Public Health, 11*(6), S11. Retrieved February 23, 2015 from http://www.biomedcentral.com/1471–2458/11/S6/S11

Dillon, G., Hussain, R., Loxton, D., & Rahman, S. (2013). Mental and physical health and intimate partner violence against women: A review of the literature. *International Journal of Family Medicine*. Retrieved February 25, 2015 from http://dx.doi.org/10.1155/2013/313909

Iyengar, S., Naik, S., Naik, A., & Poojar, M. (2013). Women in sex work: Identities and aspirations. In S. Iyengar & D. Sarma (Eds.), *A manual for counselling women in sex work* (pp. 49–64). Mumbai: Tata Institute of Social Sciences.

Jesani, A., Kapoor, A., Sagade, J, Gupte, M., Nagaraj, P., Pathare, S., & Vindhya, U. (2012). *Ethical guidelines for counseling women facing domestic violence*. Mumbai: Cehat. Retrieved September 4, 2014 from www.cehat.org/go/Dilaasa/Publications

Jewkes, R. (2002). Intimate partner violence: causes and prevention. *The Lancet, 359*(9315), 1423–1429. Retrieved February 20, 2015 from www.thelancet.com

Karandikar, S.A. (2008). *Gender-based Violence among Female Sex Workers of Kamathipura, Mumbai, India: A Contextual Analysis*. ProQuest. Retrieved February 22, 2015 from https://books.google.co.in/books?isbn=0549947744

Karandikar, S. A, Frost, C., & Gezinski, L. (2014) Patriarchy and gender-based violence: experiences of female sex workers in India, *International Journal of Social Work, 1*(1), 1–14. Retrieved February 22, 2015 from http://dx.doi.org/10.5296/ijsw.v1i1.5011

Mahapatra, B., Battala, M., Porwal, A., & Saggurti, N. (2014). Non-disclosure of violence among female sex workers: evidence from a large scale cross-sectional survey in India. *PloS ONE, 9*(5), e98321. Retrieved February 23, 2015 from http://journals.plos.org/plosone/article?id=10.1371/journal.pone.0098321

Maher, J., Pickering, S., & Gerard, A., (2013). *Sex work: Labour, mobility and sexual services*. Routledge. Retrieved February 26, 2013 from https://books.google.co.in/books

Mahila, K. (2013). *A study of situation and needs assessment of women in sex work in Uttara Kannada District* (Unpublished Manuscript). Mahila Kranti, Uttara Kannada.

National Aids Control Organisation (2013). Annual Report 2012–2013. Retrieved from http://www.naco.gov.in/upload/Publication/Annual%20Report/Annual%20report%202012-13_English.pdf

Orchard, T. (2007). In this life: the impact of gender and tradition on sexuality and relationships for Devadasi sex workers in rural India. *Sexuality & Culture, 11*(1), 3–27. Retrieved February 25, 2015 from www.researchgate.net/

Panchanadeswaran, S., Sethulakshmi., J. C., Sivaram, S., Srikrishnan, A. K., & Celentano, D. (2008). Intimate partner violence is as important as client violence in increasing street-based female sex workers' vulnerability to HIV in India. *International Journal of Drug Policy, 19*(2), 106–112. Retrieved August 6, 2014 from http://www.ncbi.nlm.nih.gov/pmc/articles/PMC2423812/

Panchanadeswaran, S., Johnson, S. C., Sivaram, S., Srikrishnan, A. K., Zelaya, C., Solomon, S., Go, V. F., & Celentano, D. (2010). A descriptive profile of abused female sex workers in India. *Journal of Health Population and Nutrition, 28*(3), 211–220. Retrieved February 23, 2015 from http://www.ncbi.nlm.nih.gov/pmc/articles/PMC2980885/#B48

Parsons, J. T. (2014). *Contemporary research on sex work routledge*. Retrieved February 26, 2015 from https://books.google.co.in/books

Peplau, L. A. (2003). Human sexuality: How do men and women differ? *Current Directions in Psychological Science, 12*(2), 37–40.

Reza-Paul, S., Lorway, R., O'Brien, N., Lazarus, L., Jain, J., … Steen, R. (2012). Sex worker-led structural interventions in India: a case study on addressing violence in HIV prevention through the Ashodaya samithi collective in Mysore. *Indian Journal of Medical Research, 135*(1), 98–106). Retrieved September 15, 2014 from http://www.ncbi.nlm.nih.gov/pubmed/22382190

Sanders, T., O'Neill, M., & Pitcher, J. (2009). *Prostitution: Sex work*. Policy and Politics: Sage Publications.

Satyanarayana, V. A., & Chandra, P. S. (2009). Should mental health assessments be integral to domestic violence research. *Indian Journal of Medical Ethics, 6*(1), 15–18.

Somanath, R. P., Mishra, R. M., Saggurti, N. & Parimi, P. (2013). The association between non-commercial partnerships and risk of HIV among female sex workers: Evidences from a cross-sectional behavioral and biological survey in Southern India. *AIDS Research and Treatment*. Article ID 108630. http://dx.doi.org/10.1155/2013/108630

Warr, D. J., & Pyett, P. M. (1999). Difficult relations: Sex work, love and intimacy. *Sociology of Health & Illness, 21*(3), 290–309.

Chapter 8
Working with Survivors of Sexual Violence: Ethical Issues and Dilemmas

Anuradha Kapoor and Chandana Baksi

Abstract This chapter examines the ethical dilemmas faced by a feminist organization while working with women survivors of sexual violence, shares principles and strategies it has developed to deal with such cases and poses questions on how to reconcile feminist principles and organizational ethics with women's agency and needs. Case studies are used to demonstrate the organization's struggles to resolve emergent conflicts between complex legal, social and psychological considerations. In this process of respecting women's autonomy, often feminist ideals and principles like social justice end up being sacrificed. The chapter also highlights the complex issues that arise in a case of sexual assault where the woman does not fit the 'victim' framework, exercises agency and has the support of multiple individuals and organizations. It emphasizes how each case is unique and multifaceted and that answers need to be found based on the specifics of each case.

Keywords Ethics · Sexual violence · Autonomy · Feminist principles

Violence against women is widespread and includes physical, sexual, psychological or emotional abuse and economic deprivation. Sex-selective abortions, female infanticide, sexual harassment, rape, domestic violence, trafficking, violence due to sexual orientation, widowhood, old age, disability and HIV infection and rape during communal and caste conflicts are amongst the varied forms of violence women and girls face, both within their homes and outside. Violence against women is on the increase and in the 10 years from 2005–2014, reported cases of sexual assault and marital violence against women have more than doubled (National Crime Records Bureau 2014).

A. Kapoor (✉)
Swayam, Kolkata, India
e-mail: anu166@hotmail.com; swayam@cal.vsnl.net.in

C. Baksi
Samikshani and Swayam, Kolkata, India
e-mail: chandanabaksi@yahoo.com

P. Bhola and A. Raguram (eds.), *Ethical Issues in Counselling and Psychotherapy Practice*, DOI 10.1007/978-981-10-1808-4_8

Swayam is a feminist organization based in Kolkata committed to advancing women's rights and ending violence against women. Swayam has been working with women survivors of violence for the past 21 years. It provides holistic support services ranging from counselling, psychotherapy, drop-in centres, legal advice and aid, follow-up with the police and referrals for vocational training, employment and shelter to survivors and their children. Swayam also conducts a variety of group interventions aimed at encouraging women to move from being survivors to active agents of social change. Through our Initiatives for Social Action and Change, Swayam tries to create an environment where violence against women itself does not take place through diverse processes. Swayam generates discussion and debate, advocates for change at the policy level, networks with organisations with similar goals, produces and shares information on women's rights issues, campaigns and mobilizes public and state action to end violence against women.

The first point of contact for a woman survivor of violence who approaches Swayam for support is her caseworker, who has basic counselling skills and para-legal training. We provide on-the-job training and supervision to caseworkers for the first 6 months, where they sit in on cases with senior caseworkers. Regular weekly meetings are held to help caseworkers address any issues and difficulties they may face whilst handling cases. Case handling, counselling and legal training are also provided. Refresher trainings are organized for longtime caseworkers and all caseworkers have access to an in-house mental health professional at a one-to-one level to deal with stress and burn-out. There are also regular group sessions on personal growth to address and resolve interpersonal issues. We have also established basic principles of casework and norms of behaviour for caseworkers to follow in their interactions with women and co-workers.

The caseworker is the primary holder of the case and provides the woman with emotional support, discusses the different options she can exercise, explains the legal rights available to her, assesses her needs, documents case records and coordinates all the services the woman may decide to access. She also networks with other stakeholders, since it is rare for any one organization to have the resources to meet all the needs of a woman. The needs assessment is done in consultation with the woman, in a non-directive manner, following which she is referred to the in-house therapist, lawyer and/or career counsellor, to the police, the protection officer, to a shelter home or for vocational training or employment, depending on the course of action she wishes to take. The caseworker also follows up the case with all the stakeholders to whom she has referred the woman and accompanies the woman to the police station and court until the woman is confident to do so herself. In case the woman so desires, the caseworker keeps in touch with her family and provides family counselling to help build support for her.

The therapist is an in-house certified mental health practitioner who is also associated with a licensed mental health organization in the city where she works with a team of therapists and psychiatrists. She offers psychological counselling to the woman and makes an objective assessment for a psychiatric referral. In cases where a psychiatric consultation is necessary, she usually works together with the psychiatrist, keeping the process transparent to the woman. In other cases, she continues to offer psychological counselling to women who are willing co-participants of the process.

Counselling, whether done by the caseworker or by the therapist, involves parallel processes that need to be dealt with in order to be ethical and effective. Issues of transference and counter-transference of caseworkers, as well as conflicts arising out of differences in the personal values of the caseworkers and the organization's philosophy are dealt with individually in the mandatory personal counselling sessions of the caseworkers and collectively in the personal growth sessions with the casework team. So far as the therapist's counter-transference issues are concerned, she is trained to deal with it herself or has the option of availing peer counselling within or outside the organization while keeping the case details confidential.

Swayam as an organization is a firm believer of transparency, social justice and equality. It adheres to the ethical principles of beneficence, fidelity, non-maleficence, autonomy, justice and societal interest in its work with women. It is here that the organization's values are often challenged when there is a conflict between two or more principles or when there is a clash of perspectives amongst activism, mental health and legal considerations.

The laws that deal with sexual violence against women and children in India include the Protection of Children from Sexual Offences Act 2012 (POCSO; Ministry of Women and Child Development 2012), the Sexual Harassment of Women at Workplace (Prevention, Prohibition and Redressal) Act (Ministry of Law and Justice 2013a), and the Criminal Law Amendment Act (CLAA; Ministry of Law & Justice 2013b). The Criminal Law Amendment Act tried to address the lacunae in the laws and procedures related to sexual abuse and rape of women and added sections to recognize other forms of violence like acid attacks, sexual harassment, disrobing a woman, voyeurism, stalking and trafficking. It also increased the penalties on the existing sections that dealt with sexual abuse of women. It defined consent as an unequivocal voluntary agreement by words, gestures or any form of verbal or non-verbal communication, communicating a woman's willingness to participate in a specific sexual act and stated that the mere fact that a woman did not resist cannot be taken as consent to the sexual activity. Marital rape, however, was not recognized in the Act except if the wife is less than 15 years of age or if she is living separately under decree of separation or otherwise. The Act has also increased the age of statutory rape from 16 to 18 years, thereby criminalizing even consensual sex between young people. Similarly, POCSO also introduced mandatory reporting of sexual abuse of children below the age of 18 years, thereby placing a legal obligation on organizations to report such cases irrespective of the child's wishes. Consequently, such legal provisions create ethical dilemmas for feminist organizations like Swayam that keep women's autonomy at the core of their work and ensure that any intervention in a case is always with the consent of the woman concerned. No decisions are imposed on her and her decision is respected and accepted even if it goes against the organization's judgment.

When a survivor of sexual abuse approaches an organization like Swayam, or is brought to a similar organization by her caretakers (in case she is a minor or incapacitated), it may be assumed that the intent of the survivor or her caretakers/guardians is to fight for justice and punish the offender, as well as to receive emotional and social support. However, while the legal component in handling these cases is of utmost importance, there are times when the legal requirement of the case is in conflict with the psycho-social needs of the survivor. In fact, the conflict or complexity of such cases is not limited to this. One of the basic principles underlying our work is autonomy, the belief that a woman has the right and the ability to take decisions about her life, and our role is to facilitate this decision-making process. Sometimes, however, the decision a survivor takes may put her life at risk and, sometimes, may also put her children at risk. This creates a dilemma for us as we are caught between supporting the autonomy of a woman, the reality of her situation and the needs of her children. There are occasions where mental health concerns of a survivor run contrary to the activist mission for social justice. Sometimes, decisions made by survivors are totally inconsistent with the values and principles of the organization. There are times when the safety of both the survivor and the caseworkers are at risk, and the organization has to ensure the safety of both parties. This is a difficult balance to achieve and we do not always have the answers. So working with survivors of sexual abuse often entails a tightrope walk between the tripartite pillars of legal, social and psychological considerations. A discussion of a few cases (see Case 8.1) can help bring to the forefront some of the complexities that we have experienced in handling cases of sexual abuse.

Case 8.1

Seema* was an undergraduate student in a well-known residential city institution. She was sexually assaulted by one of her male classmates with the help of another girl in the same institution. During a picnic where the three of them had gone to the outskirts of the city, Seema was coaxed into a room, tied and stripped. The other young woman, following the instruction of the man, let loose a couple of dogs in the room. Seema, phobic of dogs, was petrified by the whole situation. The man took pictures of Seema, hurled abuses at her and threatened to expose her photographs in public if she ever tried to 'act smart'. Seema had to beg for mercy before she was released.

Seema's elder sister brought her to Swayam a couple of weeks after the incident where she agreed to consult Swayam's in-house counsellor alongside pursuing the legal options for which she had primarily come to Swayam. Swayam's caseworkers facilitated her legal pursuit against her offender and his accomplice with suggestions from Swayam's consultant lawyer.

*Note: *All names in the cases have been changed to prevent identification.*

According to the counsellor's assessment, Seema was suffering from post-traumatic stress disorder (PTSD), a diagnosis based on complaints of sleep disturbance, nightmares, flashbacks and social withdrawal. She was referred to a psychiatrist and her counselling continued alongside. Initially, Seema was very matter of fact about the whole incident, not betraying the least hint of the feelings she had been carrying in her chest. While medication freed her of some of her symptoms, counselling sessions helped her to slowly open up with her feelings. As therapy progressed, she traced some of the feelings of helplessness and unfairness evoked by the incident to a childhood experience of sexual abuse. She shared an incident of sexual abuse experienced as an 8-year-old child. Her maternal uncle was the offender and her mother had completely minimized and even dismissed the whole incident. She recounted how she was forced to sleep with the same uncle whenever he would visit them, notwithstanding her protests and reluctance. Seema's therapy sessions unfurled a whole new journey of her life, reviewing her relationship with her mother, her self-concept and worldview.

In the process, she also re-evaluated her most recent experience of abuse and found it less overpowering. This was probably caused by two parallel, yet simultaneous processes. On the one hand, the surfacing of her childhood trauma and working on it helped her to see the two incidents as separate and her responses were no longer fused into one. On the other hand, the therapeutic process helped her to identify the deep sense of inadequacy and feelings of rejection and humiliation that lay behind her classmate's severe aggression against her. While, as an activist, it was important to ensure that the abuse was not justified, it was imperative for the therapeutic process to facilitate an understanding of the abuser's context.

Seema's focus, at this stage, shifted towards the incident of child sexual abuse, and considerable time was spent on working on her sense of guilt. She found it difficult to reconcile with her mother's attitude towards the incident and got more interested in sorting the matter out with her mother. At one point, Seema declared that she would like to go back to the ancestral home where her parents lived. She visibly lost interest in the legal proceedings, invested her lawyers with her case and left it to them to look after the court proceedings as best they could. The possible repercussions of her detachment from the case were explained to her but since client autonomy is a fundamental value respected by Swayam, her decision to move on was sacrosanct. Swayam's services, she had us understand, would no longer be necessary, and the agenda of punishing the offenders—her male classmate and his accomplice—took a backseat.

Thus, client autonomy was respected without any compromise, albeit at the cost of social justice. The organization felt frustrated that the perpetrators would get away scot-free, since Seema was no longer interested in pursuing the legal case. While at the individual level, beneficence or best interest of the client was ensured, the sacrifice of social justice probably gave a fillip to breeding offenders for the future.

The question of client autonomy giving rise to severe ethical conflict was more acutely evident in the case vignette of the Mahato girls (Case 8.2).

Case 8.2

Nitu Mahato, a 40-year-old mother of five daughters, came to the centre in extreme distress. She had long left her job in a school in order to keep an eye on her husband. She also had to keep changing her house help and residences to prevent her husband's compulsive sexual advances towards domestic workers and other underprivileged women in the localities they would move to.

Of late, Nitu had been noticing her husband's sexual interest around their daughters, aged between 8 and 15 years. She caught him peeping through the washroom door, feeling up their undergarments or hugging them tightly when she turned her back. Her eldest daughter was the first to point these out to her and Nitu's subsequent observation confirmed these allegations. When Nitu tried to warn her husband, he threatened to stop all monetary support and to abandon the family. He had previously done this a couple of years ago when Nitu had tried to protest against his sexual transgressions. Interventions by extended family members, together with Nitu's pleading and threats to inform his office, led to a psychiatric consultation. He was diagnosed with hypersexual disorder and medication was prescribed.

However, Nitu's husband stopped his medication after 3–4 months and his sexual behaviours resurfaced, accompanied by threats and verbal abuse towards Nitu.

Nitu came to us primarily with a request to save her daughters and to discuss her options around her abusive marriage. She was vehemently opposed to taking any legal action against her husband as it would cost him his job in a semi-government setup. Also, she repeatedly told us that she did not want to lose him as she was very attached to him, notwithstanding his pathologies.

After a couple of sessions with Nitu and her daughters and a visit to their current residence, we observed that all her daughters except the second one, by their own admission, were scared of their father and felt uncomfortable in his presence. The second daughter was close to the father, and instead complained against her mother's frequent emotional outbursts and abuses against family members. Nitu was close to the eldest daughter and her husband accused them of ganging up against him.

Our assessment:

(i) Nitu's daughters felt vulnerable and were unsafe in their own home.
(ii) The second daughter was most vulnerable to child sexual abuse and was probably a survivor of incest.
(iii) Nitu was suffering from severe stress that was impacting her household activities and interpersonal relations within her family, rendering her isolated and further aggravating her stress.
(iv) The family was completely dysfunctional with fusion of sub-units where the eldest daughter and mother formed one unit and the second daughter and father formed another, while the other daughters formed their own sub-unit.

To begin with, we came up with a safety plan through discussion with Nitu. We encouraged and supported Nitu to repair the bedroom and washroom doors and replace the bolts. It was decided that she would sleep with her daughters and keep the bedroom door locked at night. Strategies of ensuring that her daughters were never alone with their father were mapped. The daughters were provided with age-appropriate sex education and information about personal safety without directly implicating their father.

We tried to explore the possibility of financial self-sufficiency for Nitu. She was not convinced of its viability and was opposed to the idea from the very beginning. We further discussed with Nitu the option of sending her daughters to a low-cost or free-of-cost boarding school so that their safety from their father could be ensured. Nitu expressed her relief and gratitude and agreed to begin the necessary paperwork for admission of the girls to identified schools the following week.

However, the following week, Nitu tearfully approached Swayam, apologizing for not being able to part with her daughters or her husband. Nitu withdrew from Swayam thereafter. Initially, our follow-up calls were met with lukewarm response and eventually she stopped responding to them.

We were in a dilemma on how to handle the situation. Nitu, being our primary client and the caregiver of her daughters, could not be forced to institutionalize her daughters. Although her decision had a negative impact on her children, we were unable to decide whether we should report the case to the authorities (the regulation of mandatory reporting under the Protection of Children from Sexual Offences Act, 2012 had not been introduced at that time). We knew that if we reported the case, the girls would be removed from their home and school and sent to an unfamiliar government home, away from their mother to whom they were attached. There was also no guarantee that they would be safe in a state home. In their own home, at least, we could hope that their mother could take precautions to ensure they were protected. At the same time, we knew that the safety plan was not fool-proof and the children were highly vulnerable to sexual abuse by their father. Nitu was enmeshed in her own financial and psychological dependence on her husband and daughters and was blinded by her own needs, placing them above the basic need for safety for her daughters. In this case, the principle of autonomy of a woman to take decisions about her life as well as the lack of viable options prevented us from taking action against Nitu's will.

Thus, the right to autonomy of an individual prevailed over basic child rights to safety. Counselling failed to help her take a rational decision based on a cost–benefit analysis of her situation. Certain other questions continued to haunt us. The daughters were not safe with their father, no doubt, but would their mother be able to ensure their safety? In the absence of safe and child-friendly government-run shelter homes, what options do we have to ensure safety for these children? POCSO was not operational in this case but even if it were, could we, as a feminist organization, report this case not withstanding client autonomy? It is evident that unless the State intervenes in a concrete manner, provides sufficient resources, quality services and facilities, this problem will remain unsolved.

A different kind of conflict is inherent in Fatimah's case (Case 8.3), where the counsellor cannot but transcend the boundaries of her profession and at the same time feel short on resources and inadequate to meet the ethical standards of beneficence and non-maleficence.

Case 8.3

Fatimah came from an underprivileged family and community, living on the fringes of a metropolitan city. She was 23 years of age, married for 4 years and the mother of two small children. Fatimah's husband drove an auto-rickshaw (three wheeler) for a livelihood and her parents were very poor. When Fatimah came to Swayam's community office, she looked very ill, anaemic, walked with a limp and was very depressed. She also looked very scared and was in a hurry to leave lest her husband, who sat on guard outside our office, was inconvenienced by any delay. Her husband did not let her go anywhere alone. She had convinced him to bring her to Swayam on the pretext of enquiring about some government schemes and vocational opportunities.

Her husband bore the entire financial responsibility for the family. He was of a suspicious nature and was frequently physically and verbally abusive with Fatimah. He made excessive sexual demands on her, notwithstanding illnesses, menstrual discomfort, or the presence of children in the same room. He would beat her to compliance if she protested. From dawn till the time he left for work, he would be very attentive and extra-caring towards Fatimah, pulling up the sheets and making her a morning cup of tea but warning her not to leave the house under any circumstances whilst he was away. He would also claim that he loved her very much and all his actions, from his reprimands and battering to the sexual 'games' were expressions and proof of his affection for her.

Ashamed as she felt dirty, angry as her wishes were ignored, helpless as she felt trapped in the whole situation; she felt humiliated in more ways than one and confused with the duality in her husband's behaviour. Over the course of the counselling session she reported that she was losing interest in life and could not carry on like this anymore.

The counsellor had to strategize amidst difficult odds. Fatimah's parents, could not afford to offer her and her two children any material support or shelter. Swayam could have referred her to a shelter but Fatimah did not have the inclination, strength or energy to settle in an unknown shelter in an unfamiliar environment outside her community. Swayam could not provide her with the kind of practical support that might have been acceptable to her, for example, finding her a rented accommodation in her locality and meeting all her other financial needs.

The first (and the last) counselling session that she came for was inadequate for any long-term healing and/or safety intervention. The foci of therapy were primarily two; one, to offer her emotional support so that she could vent her anger thus alleviating some of her distress; two, to give her information about different personal preferences for physical pleasure and explain that there was nothing shameful about it if performed by consenting adults. Nothing much could be discussed about her safety in terms of exploring strategies for negotiation with her husband and she was closed to any discussion on relocation, reporting to the police or severing ties. She left the counselling space after about a 45-minute session only to come back after ten minutes howling in pain and agony. Apparently, her husband who was sitting on guard outside Swayam's community office attacked her as she stepped out of the office space, accusing her of having sex with someone behind closed doors (of the counselling room). He had apparently come into Swayam's community office a couple of times within the span of the session and tried to enter the counselling room only to find it locked. Swayam's community staff intervened at Fatimah's request to convince her husband that she was not with any man in the counselling room to prevent Fatimah being further assaulted.

Thus, started our interaction with Fatimah's husband. We spoke to him, explaining Swayam's work and clarifying the way we work with women. We took care not to antagonize him in any way as we were sure that he would take it out on his wife if he went back feeling angry or humiliated. In fact, as a strategy we invited him to all of Swayam's upcoming events in the community and subsequently gave him special attention and importance. Our intent was very clear; we wanted to have access to Fatimah and ensure that our access was not blocked in any way.

Subsequently, although Fatimah seldom visited Swayam, our community staff members visited Fatimah regularly. As part of the strategy co-evolved by the counsellor, they listened to her grievances, urged her to join the various community-based programmes, reiterated the notion of different personal preferences for pleasure and shared strategies for managing pain with the use of lubricants. On one occasion, a family counselling session was held with her and her husband with the objective of lessening the interpersonal conflict between the two.

The role that we chose to play in this case was restorative. We had a hunch that her husband was suffering from some personality disorder and we resorted to persuading him to ensure our access to Fatimah and also for her safety.

Our strategy yielded some results. Fatimah, in our subsequent home visits, would smile once in a while and definitely looked healthier. Her husband joined in most of Swayam's community-based programmes and took active interest in them, bringing Fatimah along at times. Fatimah, at one point, stopped complaining against her husband's sexual exploitation and definitely looked less depressed. Thus, our strategy seemed to bring about some positive changes in Fatimah's quality of life, though the basic power dynamics remained intact and probably more stably under her husband's control than ever before.

Here, the ethics of responsibility gave way to the ethics of intent where practical considerations reigned above feminist ideals and principles. The offender in question had criminal records and we, as a feminist organization, worked with him for the sake of beneficence for our client. This is an ongoing case and collaboration with the perpetrator is a short-term strategy we have adopted. However, this, if pursued without additional strategies to counter and balance off the perpetrator's power, would be maleficent or harmful for Fatimah. Also, the community workers dealing with the case were themselves working under a lot of risk to their lives as the perpetrator carried a gun and had poor control over his impulses. Under these circumstances, could we, as a feminist organization, stop dealing with the case and refuse to help Fatimah? What kind of safety measures can we ensure for our workers in the community which, in this case, was particularly conservative and closed to outsiders?

Finally, we would like to highlight the complex issues that arise in a case of sexual assault where the woman does not fit the 'victim' framework, exercises agency and has multiple agencies and individuals supporting her (see Case 8.4).

Case 8.4

Rita was gang-raped at gunpoint and was too traumatized to make a complaint immediately after the incident. She decided to register a complaint three days after the incident, after discussion and support from her family members. She went to a government hospital for a medical examination, where the doctor conducted a cursory examination although she told him that she had been sexually assaulted. The same evening, she went to the police station with her uncle and met the officer on duty who asked her to narrate the incident to him. She was then questioned by other male officers, including the officer-in-charge of the police station who asked her embarrassing questions about the way she was raped and the position she was in, etc., before a case was finally registered.

The investigating officer was very unsupportive and kept telling her that there was no hope for her as evidence in her case was very weak, making her extremely depressed. All this while she had not been given a copy of her FIR and when she went to collect it, a senior officer started making lewd comments. She was shocked, hurt and frightened as she knew they were making fun of the whole incident. It was then that she approached the media to get the police to take the case seriously. After the media intervened she started to receive threat calls asking her to take back her case or else her teenage daughters would be kidnapped and would face the same fate as her. Political leaders began questioning her character and said she had fabricated the case. Finally, after she was interrogated for hours by senior officers, some of those who had committed the crime were arrested.

Her medical examination was conducted more than a week after the rape due to delay by the police and the medical establishment. For the medical examination, she was made to undress with the windows wide open, and stand with her legs apart and hands stretched out. Although the doctors comforted her, reassuring her that the accused would be punished once they were caught, they conducted their examination in a very clinical manner, poking and prodding her without realizing how she felt.

When we met Rita, she was extremely emotionally distraught and it was clear that she was under severe mental trauma. She was on sleeping pills as she was unable to sleep properly, had regular nightmares and any sound on the street would keep her awake. She was frightened as she was receiving threat calls and was really worried about the safety of her children. She had scars on her face and her lip was still swollen. She had police protection for herself and her children and the media were all parked outside her house. We offered her our support for counselling, her legal case, help with following up her case with the police and any other support that she required.

Over two months, Rita spoke to Swayam's caseworker regularly on the phone—she reached out for emotional support and spoke about what was happening to her, her feelings of helplessness and vulnerability. She talked about how her neighbours and landlord were pressurizing her to move out of the locality. She was afraid for her school-going daughters and for her own safety as she continued to get phone calls threatening her to take her case back. All of these experiences further compounded the trauma that she was going through.

After two months, Rita came to Swayam's centre and sat with the therapist as well as her caseworker. We offered her continued therapy and legal support in the form of a de facto lawyer. She said she did not want a lawyer at the time and later informed us that she already had a de facto lawyer through another NGO. Though she initially said she would come for therapy, she kept telling her caseworker that she felt good talking to her and did not feel the need to go for psychological counselling.

On the legal front, when the charge sheet was filed and the date was set for framing charges, the accused applied for bail in the Supreme Court. She was very worried as one of the organizations that was supposedly helping her told her that the accused were likely to get bail. We intervened, helped her get legal representation in the Supreme Court and the bail was denied.

The trial of the case began more than a year after the incident and she was further traumatized by the court process. She was in a roomful of male lawyers, representing each of the three accused where only the judge, public prosecutor (PP) and clerk were women. She was scared to go to the court as the family members of the accused would be there and could intimidate her. Although the trial was in camera she did not have any support persons who could go into the courtroom with her. We explained to her that she had a right to have support persons in the courtroom, and that we would be able to accompany her if she made an application to the court asking for the same. However, her lawyer and the public prosecutor told her that this would delay the case. She feared that they would get angry and offended if she insisted. At the best of times the public prosecutor would be angry with her as she defied the image of being a 'victim'.

During the trial, the first seven to eight dates went by with the defence lawyers asking her questions that bore no relevance to the case but made insinuations about her lifestyle. The judge, whose duty it was to create a conducive atmosphere for a rape victim to testify in court and to protect her from unnecessary and humiliating questioning, did nothing. Not only was the judge not empathetic, she made it clear she did not like Rita. Despite Rita being asked embarrassing questions quite unrelated to the case, the judge did not stop this line of questioning nor did her PP or de facto lawyer object.

She shared how humiliated she felt when her clothes and undergarments were being displayed in the courtroom and the lawyers were laughing and the accused were smirking and the judge did not intervene. Her evidence and cross-examination were long drawn-out and took their toll on her emotionally. After each day in court, she would be angry, depressed and upset. None of the protocols laid down under the law were complied with by the police, medical establishment or the court, and she was consistently denied the support that a victim of sexual assault was entitled to throughout the criminal justice process, which was long drawn-out and traumatic.

On the one hand, Rita was strong, vocal, fought actively for her rights and exercised agency and control over her case. She did not fit the victim prototype and was very upfront about her right to live her life on her own terms and not be dictated by societal norms. As a result, she was constantly rebuked, reprimanded and judged by all the state agencies she came into contact with as well as society. At the same time, her candour and courage also attracted multiple organizations and individuals to help and support her. Consequently, she had too many advisors giving her differing advice. Whilst she had many people backing her and she talked to different people and organizations openly, there were very few people she actually trusted and she did not want to depend or rely on anyone except herself.

On the other hand, she was very vulnerable and would seek emotional support and solace from her caseworker. She shared how some people who had helped her, were taking advantage of her to get personal mileage/publicity, and were pressurizing her to take decisions that were not in her best interest, saying she was obligated to them. The caseworker could see that while she was daring and fighting for justice, behind her external façade, her trauma was intensified by her experience with the criminal legal system and pressure from some of her so-called advisors and well-wishers.

During the course of our working with her, numerous people approached us to get access to her to ask her to do things that may or may not necessarily have been in her best interests. Our organizational policy does not allow us to give anyone direct access to the survivor without her consent. However, the organization was in a dilemma about whether or not to convey information that we felt was not in her best interest. The principle of agency demanded that we passed on the information to her and let her make her own decision, yet we were aware that she was under tremendous pressure and was susceptible to being taken advantage of. Ultimately, the principle of autonomy overrode these concerns but we would discuss the pros and cons of each choice with her so that she could base her decisions on an informed understanding of the situation. We encouraged opportunities that arose for her to share her experience in safe spaces that would lead to an improved understanding of the trauma and lack of support that rape victims experience within the criminal justice system, and to influence people within the system to address these issues.

Her healing process was somewhat helped by her transition from voicing her personal experience to supporting other women who had been sexually assaulted and becoming an advocate for women's rights in these cases. However, this too was fraught with a dichotomy. Whilst she wanted to be an advocate, she did not always have the autonomy to choose when, where and how she would advocate. She talked about how she was forced into certain situations of advocacy that she was not be comfortable or ready to deal with. Understandably, she still had unresolved trauma-related issues, and such situations generated further emotional trauma for her which she found very painful.

While she was strong outwardly, her internal trauma was not visible. She was in a state of emotional turmoil and needed professional help to deal with her trauma. However, she ignored this suggestion each time her caseworker brought it up. She said she was happy with the counselling provided by her caseworker and did not agree to see a professional therapist despite constant encouragement from her caseworker. She needed a job and wanted a backend job where she did not have to come face to face with people. We tried unsuccessfully to help her in this area. Time and again, we offered the services of a de facto lawyer who would stand up for her rights in court, protest unnecessary questioning and petition for her to be allowed support persons in court. However, she did not take up our offer as she was under pressure from some of her advisors to engage specific lawyers who did not allow us any access to them.

She trusted and connected with her caseworker at a personal level but was somehow not very responsive to support from the organization as a whole. However, her personal equation with the caseworker complicated the organizational dealings with her—each time the organizational support was offered, she retreated. Whilst the organization felt that it could do a lot more for her, it was unable to because Rita was not seeking professional help of any sort, but reaching out to the caseworker personally in situations of crisis to the exclusion of the organization. As the organization was one of the many who extended support to Rita and organizational attempts to reach out were, more often than not, frustrated, the case could not be pursued systematically by the organization as a whole. Moreover, organizational policy prevented the caseworker from connecting with the victim/survivor at a personal and social level. This is based on the understanding that if a caseworker gets personally involved then the optimum professional detachment that is required to assess and support the case may become biased and endanger the survivor's best interest. However, realizing that Rita was in a crisis situation and reaching out for support, albeit at a more individual level, the organization took the decision of extending support to Rita on her own terms and providing support whenever possible.

Whenever Rita communicated dates to them, the caseworker and others from the organization also accompanied her to court, waited with her outside the courtroom till the case began and stayed till she came out after her deposition. However, there was no consistency in her communication with her caseworker. At times she would be in constant touch and at other times there would be no communication. She made appointments with the counsellor/caseworker on numerous occasions but did not keep them. Whenever Rita reached out, the caseworker provided her with emotional support, a safe space to vent and be understood without being judged and acted as a trusted ally to discuss the positive and negative sides of the various proposals she was offered. But the caseworker often felt demotivated and frustrated as she could not follow-up Rita's case systematically due to the breaks in communication, her being wary of organizational support, and the limitations placed on providing her support in the legal process. All this limited the extent to which the caseworker or organization could provide support to her in terms of advocacy within the criminal justice system, claiming compensation, victim protection, employment, etc. Consequently, the caseworker often felt that she was not doing enough.

The fact that Rita did not fit the image of a victim resulted in both positive and negative consequences and put her under tremendous pressure. She was in the midst of a tug-of-war between the different players in her case. The organization in this case felt that there was a conflict between the principles of autonomy and beneficence as the exercise of autonomy did not always serve the best interest of the client. Further, the personal connect with the caseworker and wariness to engage regularly with the organization resulted in frustration at the level of the organization and the caseworker to effectively pursue the case. Ultimately, Rita did get justice from the court and the accused were convicted. However, the process of getting justice was as important as the conviction and it is here that Rita was let down at every step by the criminal justice system.

The four cases discussed above deal with sexual abuse committed by a peer, a father, an intimate partner and a group of strangers, respectively. In three out of the four cases, the survivor of the abuse was at the helm of decision-making with regard to the handling of the case and in the fourth, the mother, as the guardian, was the decision maker. Although the personal is political in the realm of social activism, sexual abuse from the perspective of the survivor is a very personal experience. This, together with the often-internalized social stigma around anything sexual, makes it imperative that autonomy of the survivor is never challenged in facilitating recovery from the trauma. As a feminist activist organization we try to support the survivors in their journey towards recovery and enrich our understanding of the issue and experience in the field in the process. In all of these cases, a common thread of conflict is discernible; while the survivors or their caretaker (in the case of the minor girls) exercised autonomy, the organization felt restricted by the self-determination of the survivors (or their caretaker) and felt helpless when the survivor/caretaker took decisions that the organization believed were not the best options open to the survivor/caretaker. Thus, in all four cases the challenge is posed by survivor autonomy; limiting the scope for organizational intervention on the one hand but ensuring an empowering experience for the survivor on the other.

Besides, in the absence of safe and quality government homes for abused women/children, the lack of access to quality education from a shelter home, the reluctance of women to move away from a known community situation to an unknown shelter home, the inability of the organization to provide for women's social and economic needs, and the importance of not creating dependence but supporting autonomy; the role of the organization and its capacity to influence the decisions that survivors or caretakers make is limited.

At the end of the day, there are no easy answers in how to deal with these cases. Despite the systems, processes and policies that guide our work, we continue to struggle with the right course of action as each case is unique and presents its own peculiarities. We constantly have to review our strategies to deal with the realities and to act in the best interest of each woman. Our systems and processes are not frozen in time; they are ever evolving and we have instituted a system of reflecting on them in our regular casework meetings as well as during our annual planning sessions. As a feminist organization, we try to build consensus and give weightage to all opinions in the organization before coming to a decision on handling challenges and conflicts, and encourage the survivors we work with to share their feedback on the services they access through an annual written feedback exercise.

Despite our best efforts, there are also many questions for which we have no answers and we often feel alone in our efforts to find the best solutions. One way in which we have tried to address these ethical conflicts and dilemmas is by organizing an annual interactive session with senior activists, legal experts, therapists and psychiatrists from other organizations and institutions to discuss these critical cases and reflect on possible strategies/courses of action. Participating experts also share their own experiences and challenges they encounter in the course of their work and the forum is utilized for mutual learning and reviewing basic principles and standards of care and support for survivors. At the same time, it is imperative for us to focus on the lacunae in State services and facilities for survivors and advocate for quality services for women. While principles are useful in guiding us in our work, the reality of each case differs and we will continue to face dilemmas and grapple with what would be the best course of action under specific circumstances depending on the complexity of each case.

References

Ministry of Women and Child Development. (2012). *The Protection of Children Against Sexual Offences Bill.* Retrieved on 13.05.2016 from http://wcd.nic.in/sites/default/files/childprotection31072012.pdf.

Ministry of Law and Justice. (2013a). *Sexual Harassment of Women at Workplace (Prevention, Prohibition and Redressal) Act.* Retrieved on 13 05.2016 from http://indiacode.nic.in/acts-in-pdf/142013.pdf.

Ministry of Law and Justice. (2013b). *The Criminal Law Amendment Act.* Retrieved from http://indiacode.nic.in/acts-in-pdf/132013.pdf.

National Crime Record Bureau. (2014). Crime against Women, http://ncrb.nic.in/StatPublications/CII/CII2014/chapters/Chapter%205.pdf. Retrieved on 13th May 2016.

Chapter 9
'Coming Out' of the Comfort Zone: Challenging Heteronormativity Through Affirmative Counselling Practice with Lesbian and Gay Clients

Ketki Ranade and Shruti Chakravarty

Abstract This chapter focuses on the historically ambivalent relationship between mental health sciences and homosexuality. It briefly traces this history and then moves to discussing the need for an ethical and affirmative stance while working with lesbian and gay clients. The chapter discusses some of the common dilemmas of counsellors/mental health practitioners while working with their lesbian and gay clients.

Keywords Gay affirmative counselling practice (GACP) · Lesbian and gay (LG) clients · Gay-biased/unethical practice · Sexual minority stress

9.1 Ambivalence About Homosexuality Among Mental Health Sciences

Historically, homosexuality has been viewed as a sexual perversion within medical sciences since the early 1900s. This view is reflected in homosexuality being classified as a form of mental illness in the first Diagnostic and Statistical Manual (DSM) of the American Psychiatric Association (1952). In this first edition of the DSM, homosexuality was classified under 'sociopathic personality disturbance'. This later changed to a classification under the category 'sexual deviance' in DSM-II. Some of the treatment methods that have been historically used to treat homosexuals have included lobotomy, hypothalamotomies, implantation of testicular tissue of heterosexual men into homosexual men, induced seizures, electric

K. Ranade (✉)
School of Social Work, Tata Institute of Social Sciences, Mumbai 400088, India
e-mail: ketki.ranade@tiss.edu; ranade.ketki@gmail.com

S. Chakravarty
Tata Institute of Social Sciences, Mumbai 400088, India
e-mail: shruti.network@gmail.com

P. Bhola and A. Raguram (eds.), *Ethical Issues in Counselling and Psychotherapy Practice*, DOI 10.1007/978-981-10-1808-4_9

shock and behavioural methods such as masturbatory reconditioning, aversion therapy and so on (Haldeman 1994; Silverstein 1991, 1996).

In the early 1970s, several social movements including the African-American Civil Rights movement, the Feminist movements and finally the Gay Liberation Movements made a significant impact in challenging the medical view of homosexuality. These efforts accompanied by research studies such as the one by Hooker (1957) and earlier research studies by sex researchers such as Kinsey (1948, 1953) played a significant role in declassification of homosexuality from the DSM. While homosexuality as a diagnosis ceased to exist in the DSM, a new variant in the form of Sexual Orientation Disturbance and Ego Dystonic Homosexuality entered the DSM. It was only in the revised 3rd edition of the DSM in 1987 that homosexuality was entirely deleted from the list of mental disorders (American Psychiatric Association 1987). Other classification systems of mental illnesses such as the International Classification of Diseases (ICD) by the World Health Organization (WHO) followed this move and removed homosexuality from its classification in 1992 (WHO 1992). It is thus important to note how recent these shifts have been in mental health sciences vis-à-vis homosexuality.

In the Indian context, while most mental health practice continues to be guided by Euro-American models of mental health training and practice, the same does not seem to hold true when it comes to homosexuality. Mental health professionals (MHPs) in India have at different points of time in history (after APA's declassification of homosexuality in 1973) conducted various kinds of reparative, conversion treatments with their homosexual clients. For instance, the Indian Journal of Psychiatry carried articles in the years 1979, 1982 and 1983 about use of electrical aversion and other techniques of behaviour therapy for 'cure' and 'change' of homosexual orientation (Pradhan et al. 1982; Mehta and Deshpande 1983; Sakthivel et al. 1979). Similar practices have been documented in other, more recent studies about use of masturbatory reconditioning, aversion therapy including mild shock as well as hormonal treatments (Ranade 2015a; Kalra 2012). There has been a lack of any concerted dialogue among mental health professional bodies in India about what is the scientific position of Indian mental health professionals vis-à-vis homosexuality. Do MHPs in India accept the international position with respect to sexual diversity and affirmative practice with lesbian, gay, bisexual, transgender, and queer (LGBTQ) groups adopted by several international bodies such as the APA, WHO and so on? Studies indicate that there have been large gaps and silences about homosexuality among mental health professionals in India. Parekh's study (2003) focused on the extent to which sexual minority concerns are represented in mainstream research publications of MHPs in India. This study reveals that among 829 papers published between 1974 and 2000, in the *Indian Journal of Clinical Psychology* (an official publication of the Indian Association of Clinical Psychologists), there were only two papers on the subject of homosexuality. Similarly, there were four research papers published in the *Indian Journal of Psychiatry* between the years 1982–1995. All these papers were case studies of homosexual patients and their treatment outcomes.

The LGBT movement/s in India have been vibrant and assertive, especially in the last decade and a half in most urban and some rural parts of the country. The decriminalization of homosexuality through the Delhi High Court's Verdict on Section 377 (section that criminalizes homosexuality) of the Indian Penal Code has been one of the major boosts for the LGBT movement/s in India. After this verdict of the Delhi High Court in 2009 and the increasing visibility of LGBTQ issues in the media and the public domain thereafter, there have been a few papers in the *Indian Journal of Psychiatry* (IJP), the official publication of the Indian Association of Psychiatrists, discussing the need for depathologization of homosexuality in clinical practice (Kalra et al. 2010). In an editorial in 2012, the *IJP* clarified its position vis-à-vis homosexuality, stating that it is a normal variant of human sexuality and underscored the need for more research on LGBT lives in India (Rao and Jacob 2012).

Thus, comparable to the situation in America that led to the removal of homosexuality from the list of mental disorders, in India too, change in legal status of homosexuals and a dynamic LGBTQ rights movement has had a significant impact on attitudes of Indian MHPs. There have been more voices within the mental health fraternity who have been speaking in support of the cause of LGBT rights. In fact, a group of MHPs signed a petition to the Supreme Court of the country in the context of an appeal on the Delhi High Court Judgment of 2009 by various right-wing groups seeking to reinstate the Section 377 of IPC. MHPs stated that it was their medical opinion that homosexuality is not a form of mental illness and on the contrary, a law like the Section 377 causes tremendous psychological stress and trauma to homosexual persons.

Interestingly, in December 2013, the Supreme Court of India in response to a challenge to the 2009 judgment of the Delhi High Court, upheld Section 377 of the Indian Penal Code and in effect (re)criminalized homosexuality. Coincidentally or possibly influenced by the Supreme Court judgment, the *IJP* in January 2014 published a letter to the editor titled, "A fresh look at homosexuality", where the author while stating his sympathies to the gay rights cause also expresses his reservation in accepting homosexuality as a 'normal' sexual variant and argues in favour of the position of homosexuality as a deviance. Verghese (2014) in his letter states,

> A normal variant cannot be considered completely normal. It is, in fact, an aberration in the psychosexual development, caused by genetic and psychosocial factors for which the person is not responsible. There are research findings, which suggest that there are structural differences in the brains of people with homosexual orientation… Every biological function has a physiological goal and purpose. Sexual activity has two goals. One is procreation to safeguard the continuation of the species. The second one is the experience of pleasure, which in fact, is to facilitate the sexual activity and to strengthen the bond between husband and wife.
>
> Homosexuality negates one of the goals of sexual activity procreation. Homosexuality has therefore, to be considered as an aberration in the psychosexual development caused by genetic and psychosocial factors… (pp. 209–210).

It is amply clear, not only from the situation in India but the world over that homosexuality has been one of those controversial issues that has been dealt with and on occasions resolved not merely through what we consider as 'scientific knowledge' and 'fact' but through consideration of social/religious/moral attitudes and opinions of the medical/mental health community. The crucial questions then, in the context of ethical practice with homosexual clients are: To what extent does the clinician's attitudes towards sexuality, homosexuality in particular, affect their practice with their homosexual clients? To what extent are clinician attitudes shaped by social situations of marginalized groups at a given time period? How can clinicians become aware of their own moral and social judgments that arise from their own social positions? Some of these questions, which are at the core of ethical and affirmative practice with gay and lesbian clients, will be discussed in this chapter.

Pathologization and depathologization of homosexuality have been, at least in part, determined by social attitudes. Similarly, social attitudes towards homosexuality have a significant influence on the levels and intensity of distress experienced by gay and lesbian individuals in a given society. Sexual minority stress is a concept that has been used by researchers to refer to the psychosocial distress caused by a stigmatized sexual identity (Meyer 1995). Sexual minority stress or gay-related stress is another concept that will be discussed in detail in this chapter as it forms the base of affirmative mental health practice with gay and lesbian clients.

9.2 Need for Gay Affirmative Counselling Practice

Gay affirmative counselling practice (GACP) is not a new form or model of counselling or psychotherapy. It includes modifications in existing practice to incorporate the issues and stressors inherent in living as a sexual minority in a heterosexually constructed world (Ritter and Terndrup 2002). It includes any kind of therapeutic work that communicates the belief that homosexuality is a natural developmental outcome for numerous individuals. There has been extensive work done in the American and British context by the American Psychological Association and the British Psychological Society to establish guidelines for GACP (American Psychological Association 2000; Shaw et al. 2012). The need for GACP arises primarily out of two main reasons. First, homosexuality has been traditionally viewed as a sexual perversion and deviance and some of these ideas tend to percolate in the practice of many mental health professionals, teachers and trainees, who have been trained in these traditional paradigms. Also social attitudes towards homosexuality often tend to be negative, which can be internalized by practitioners. The second reason is that, many gay and lesbian individuals tend to experience unique stressors related to their stigmatized sexuality and this leads to sexual minority stress, which would be significant to work on in clinical situations. Hence, there is a need for mental health professionals to be trained on these unique stressors and experiences of gay and lesbian clients.

In this chapter, while we will not discuss specific tenets of GACP, we will discuss the gay affirmative stance/position with respect to several ethical issues that come up while working with gay and lesbian clients in counselling and clinical settings.[1]

9.3 Sexual Minority Stress and Gay-Related Stressors

There has been extensive work done by psychologists on the effects of marginalized/minority status on the mental health of members of minority community. Meyer (1995) describes sexual minority stress as the psychological distress that is a result of the stigma, discrimination, violence experienced by sexual minority individuals as a result of their sexuality. Minority stressors are conceptualized by Meyer as internalized homophobia that relates to gay individual's directing negative societal attitudes towards the self, stigma that relates to expectations of rejection and discrimination and actual experiences of discrimination and violence (Meyer 1995). A few defining features of minority stress are: (a) it is *unique* and hence 'additive' to the general stressors that everyone experiences and in that sense requires a response over and above the one required to cope with other life stressors; (b) it is *chronic*, implying it is related to relatively stable underlying social and cultural structures; (c) it is *socially based* and stems from social processes, institutions and structures that are beyond the individual (Meyer 2007). Some other studies point to the extra energy that is expended examining reasons for one's sexuality and figuring out one's sexual identity while living and growing up in a heterosexist world, maintaining multiple identities, stress involved in the process of coming out (publicly identifying oneself as LGBT), familial issues and a general lack of social support as some of the unique stressors related to a gay or lesbian identity (Davison 2001). Smart and Wegner (2000) describe the cognitive burden involved in hiding and living in secrecy. In the Indian context, some of the gay-related stressors that have been documented include, pressure for heterosexual marriage and forced marriages, pressure to seek help/cure from doctors to change sexual orientation, violence from the natal family (Ranade 2015a; Fernandez and Gomathy 2003; Ghosh et al. 2011; Labia 2013). Another qualitative study with forty gay and lesbian individuals indicates that some of the common gay-related stressors included a sense of isolation and absence of gay affirmative language and

---

[1]This chapter is based on learnings and insights from a study conducted by the authors titled, 'Conceptualizing Gay Affirmative Counselling Practice in India supported by Saksham, The Global Fund to Fight AIDS, Tuberculosis and Malaria, Round-7, Counselling Component as well as an e-manual written by the authors on Gay Affirmative Counselling in 2013, available at: http://www.academia.edu/22507988/Gay-Affirmative_Counselling_Practice_Resource_and_Training_Manual.

images while growing up, working out questions of 'abnormality', 'deviance', denial about one's sexuality, fear of being found out, internalized homophobia and so on (Ranade 2015b).

9.4 Ethical Concerns While Working with Gay and Lesbian Clients

This section highlights some of the common issues/concerns faced by counsellors while working with gay and lesbian clients. As mentioned in the first section of this chapter, an ambivalent relationship has existed historically between mental health sciences and homosexuality and this is often at the root of one of the most common issues faced by MHPs. MHPs often find themselves faced with clients, who find it difficult to accept their same-sex sexual attractions and are seeking for some way out of this situation of stress, shame and embarrassment. Similarly there are times, when family members or friends bring a gay or lesbian client to a therapist/MHP asking for changing their sexual orientation. In both these situations, what the MHP is hearing, loud and clear is that the client or the family/friend wants a 'way out', 'for the shame to stop', 'finding a way to live a life of dignity/self-respect' and so on. This is then interpreted as client seeking a 'cure/change' of sexual orientation by all stakeholders, i.e. clients themselves, their families and also the MHP. This section will suggest some ways forward when faced with this common dilemma in practice with LG clients.

Some of the other concerns that MHPs face includes questions such as, How much do I need to know about gay peoples' lives to be able to work with them? Is my training as a counsellor enough to do GACP? Do I need to be gay myself to be able to effectively and ethically work with gay or lesbian clients? Is it necessary for me to disclose my sexuality, whatever it maybe to my LG clients? Will that be a hindrance or a facilitating factor? We have tried to address some of these questions here.

9.5 Working with Clients Who Want to Change Their Sexual Orientation

Very often, clients are unable to accept their own sexuality and may request a change. There would be others forced by or pressurized by family to seek cure/change. When MHPs are faced with such clients, it is important that we recognize that a request for 'cure' or 'change' is a reflection of clients' inability to accept their non-heterosexual sexuality. There are invariably stressors or life events (in the immediate context or long drawn) that underlie this request. GACP would focus on these underlying reasons; in the environment as well as within the client in the form of internalized homophobia. In other words, it would be an ethical choice to respond to requests for conversion treatment by initiating a process of self-acceptance.[2]

This ethical choice is grounded in two realities. There is extensive research on conversion treatments that asks serious questions of these forms of treatment both on the grounds of efficacy and ethics (Haldeman 1994). These treatments are not only known to not work but to produce harmful effects on individuals who are subjected to the same (Smith et al. 2004). The Yogyakarta principles on application of international human rights law in relation to sexual orientation and gender identity considers any efforts at treatment, cure or suppression of sexual orientation and gender identity as medical conditions to be a form of medical abuse under Principle 18 (Yogyakarta Principles 2007). Furthermore, GACP advocates that the use or supporting the use of any method that attempts to 'change', 'cure' sexual attractions or orientation is unethical. Trying to find causes for sexual orientation is again upholding the belief that there is something that went wrong and the person deviated from the standard norm of heterosexuality. Asking sexual minorities why they are homosexual is like asking straight people what caused their heterosexuality.

However, this does not mean that MHPs send away clients who seek change/cure. That would be more damaging for the client. Also internalized homophobia is often so deeply ingrained that clients may seek out another practitioner who would be willing to 'cure' them and as we know there is no dearth of practitioners of all shades and hues, who may provide unethical treatments. So the choice is then to provide information to clients about the inefficacy of conversion treatments, while simultaneously addressing their motivations to seek change and working towards self-acceptance. In order to facilitate self-acceptance and ferret out the negative feelings clients may associate with their sexuality, the practitioner needs to know about some of the impediments to self-acceptance. These include,

Isolation Growing up in a heterosexually constructed world would mean that clients often grow up feeling that they are the 'only one like this'. They may not know that same-sex desires are normal and that there are others like them who feel the same way.

Invisibility The silence and invisibility around homosexuality means that clients may not see their lived reality reflected anywhere around them. They may not see people like them, resources for them, images that celebrate their reality, or any sort of experience that makes them believe that, 'it's possible to live like this, with dignity and respect'. Often, even the language to address themselves and their desires in a positive manner is missing from their lives.

Internalized homophobia Negative messages, stigma and rejection by society are often internalized by clients, such that they start believing these messages to be true for themselves, their sexual desires and their relationships.

[2]Also see Report of the American Psychological Association Task Force on Appropriate Therapeutic Responses to Sexual Orientation. Available at: www.apa.org/pi/lgbc/publications/.

Denial Many clients may not be able to recognize or articulate deep-seated feelings of homophobia and rejection.
Institutions Many religions and religious leaders 'ban' homosexuality deeming it unnatural. Marriage is another institution that puts huge amounts of pressure on lesbian and gay individuals. The pressure to marry and produce children to continue the name and family lineage or to continue appearing 'normal' in the eyes of society is high. Thus, institutions such as the family, marriage as an institution, law, education, religion often condemn homosexuality, which makes self-acceptance an arduous task.
Individual losses Following from the above idea that most social institutions favour heterosexuality and condemn homosexuality, clients may have to give up many benefits that come from being heterosexual if they decide to accept themselves the way they are. The loss of heterosexual privilege or not being able to live the heterosexual script may result in many individual losses, big and small.

A GACP practitioner needs to be aware of these to be able to help clients work on barriers to self-acceptance as well as understand client resistance for the same. Some of the other steps involved in working on self-acceptance include

- *Validation* Counselling spaces are often one of the first few spaces where the client is talking about his/her same-sex desires. Hence the responses that the client receives here are extremely vital in the future journey of the client. If the client experiences prejudice, discomfort or stigma then it is likely that the client's process of self-acceptance will receive a severe setback and the client may feel further driven into the closet. On the other hand, if the first experience is validating, where the client feels accepted, is assured that he/she is not alone, is not pathological, is given information on resources for gay people, then the client's process of self-acceptance is likely to be accelerated.
- *Mirroring* Mirroring is an important process in development of the 'self'. All of us need to see positive images of ourselves being reflected back to us by significant people in our lives such as our families, our friends, neighbours, communities and so on. However, most gay individuals may have missed this in their growing up years, especially with regard to areas where they may not be conforming to the norm. This can cause a lot of insult and trauma to the developing 'self'. Therefore, providing affirming, validating mirroring experiences to gay clients in counselling spaces with respect to their sexual and gender expressions is very important. This can provide the clients with an emotionally corrective experience that may be vital in self-acceptance.
- *Shift from ego-dystonicity to social-dystonicity* Helping clients' to place the onus of their problems due to a heteropatriarchal, homoprejudiced society outside of themselves onto this very society is useful. It is like saying to the client, "*That you are homosexual is not your fault, but that you are having trouble being a happy, healthy homosexual is because of the situation you are in*". In this sense the locus of the problem is not inside the clients that they have to be ashamed or embarrassed about, it is located elsewhere and clients need not feel like they are the problem.

- *Therapeutic groups* Therapeutic groups are another effective strategy to enhance self-acceptance in clients who have high degree of internalized homophobia. Group sessions help to reduce isolation that many clients may have faced while growing up. They also provide safe ventilation spaces. Listening to each other's stories of struggle and resilience can be cathartic as well as healing. Such group sessions, if conducted by a skilled counsellor, can provide for spaces for emotionally corrective experiences, cohesion and a feeling of community and support among its members.
- *Use of expert position* Psychoeducation plays an important role in counselling, in general. In this context, providing basic information about homosexuality, which is non-pathological, normalizing, affirmative and informative about resources for gay people, is very useful. Given the level of isolation of some of our LG clients and their family members, providing information on websites, pamphlets, support groups and so on can be very useful. Over and above this, counsellors, just like doctors, are seen as being in a position of authority and as subject experts. It may be useful sometimes to use this authority/power to guide LG clients and their families to affirmative and scientific information and help them avoid unscientific and unethical information about conversion treatments that maybe available in the market (see Ranade 2015a for information on use of conversion treatments by practitioners in India).
- *Awareness and exploration of sexual minority stress* Homoprejudice, stigma, discrimination, internalized homophobia and difficulties with self-acceptance is a vicious cycle. The more the number of negative life events and experiences of violence and discrimination related to one's sexuality, the more the individual is likely to be isolated, feel ashamed, inadequate and unsupported and the more the number of difficulties with self-acceptance. It is important that counsellors recognize this cycle and help their client's make the linkages between stigma and their mental health status in counselling sessions. Making these linkages explicit to gay clients and exploring the same in context of life span, i.e. childhood, adolescence and so on is an important component of GACP. For instance, several gay and lesbian individuals recount stories of ways in which their non-normative gender and sexual expression was pointed out and corrected in contexts of family, friend circle, playmates, school, college and so on. These corrections included messages about normative ways of dressing, play, mannerisms as well as acceptable sexual interests and expressions in the opposite sex. These corrective responses may range from being merely advice giving about normative behaviour to teasing, to coercion as well as violence and punishments. Exploring the emotional impact of these early corrective responses and their role in impeding self-acceptance is crucial.
- *Connecting clients to gay affirmative resources* In the context of isolation, helping clients to know more about other gay people and their lives through magazines, websites, fiction and non-fiction literature, meeting LG persons, knowing about support groups for LG persons, events including social events such as parties is all very useful in the process of self-acceptance. The role of the MHP here is twofold. Firstly, if MHPs are working with LG clients they should

collect basic information on resources for LG clients and make that available at their clinic/set-up. Secondly, the therapist must judge when the client is ready to explore this gay world. Some clients may be struggling with a lot of internalized homophobia and may have several stereotypes about LG persons. A virtual or direct connect with LG individuals maybe overwhelming for the client. Thus, this process has to be done in a phased manner checking with client about their readiness.

9.6 Educating Oneself as MHPs on LG Issues

One of the biggest challenges for MHPs in working with LG clients is that their training or life experience does not really equip them to understand the minority stress experienced by their clients or the unique life challenges that are faced by sexual minorities across their lifespan. This is especially true for heterosexual/straight counsellors. Most MHP curricula in India rarely touch upon questions of non-normative genders or sexualities. However, if MHPs want to work with LG clients in an affirmative manner, they have to choose between a homo-ignorant stance and a gay affirmative one. A homo-ignorant stance would mean a stance, where the MHP is not really aware of LG issues and deals with them just like any other issue at their clinic. This stance would typically argue that 'depression is depression', whether it is of a housewife, a successful business man, a poor homeless person on the street, a young person dealing with their non-normative sexuality. GACP would state that there is a definite link between same-sex sexuality, the stigma associated with it, the trauma and resilience involved in consolidating this identity and mental health consequences of distress or illness.

The other aspect of homo-ignorance is that, straight/heterosexual counsellors are also primarily exposed to the heterosexual model of living and hence may not know enough about gay lives, struggles, relationships, social scene or the emergent rights discourse in the context of sexual identity and gender orientation. Some counsellors may believe that they can learn from their own LG clients. However, merely relying on one's clients to educate oneself is not ethical practice. It may be useful for heterosexual/straight identifying counsellors to read about, interact with gay individuals outside of the clinic/counselling space and engage with resources available for gay individuals. For instance, counsellors could read books/watch films with gay themes or meet with gay individuals at NGOs or drop-in centres. This helps counsellors to be more educated and informed about gay lives and also to pick up more nuances about some of the issues that their clients bring up in sessions. By knowing about these different resources and engaging with these, counsellors will also be better educated about other services available for gay individuals. If LG clients know about these resources from their counsellors, it would reduce the clients' isolation. Knowing about resources and services for gay individuals would help counsellors to make better referrals and provide more effective services to their clients.

Finally, living one's life as a 'gay' person may get difficult due to the external environment that is heteronormative, which views heterosexuality as the only natural, normal, universal, moral and pure form of sexuality. This context of social marginalization forms the framework for understanding some of the common mental health concerns or unique life stressors that LG populations encounter. Gay affirmative therapists need to educate themselves on these. Being gay informed is at the crux of affirmative practice. The unique life stressors form the underlying basis for some of the mental health manifestations or problems that clients bring to sessions. It is unlikely that clients themselves may be able to pinpoint these stressors in their lives. They will tell their counsellors about feeling depressed, ashamed or dirty, they may tell you that they are not being able to concentrate on their studies or being able to hold down a job. They may express that they get teased or beaten. Making the possible link of client problems and manifestations to the unique life stressors related to their sexuality, is the responsibility of the GAC practitioner.

9.7 Self-awareness Among MHPs

Self-awareness on the part of the counsellor is vital in all counselling processes. Knowing about one's own biases, inadequacies, seeing oneself as fallible and therefore open to correction, change and growth lies at the heart of this principle of 'counsellor self- awareness'. Just as knowing about one's limitations is significant, knowing about feelings and thoughts that clients generate within the counsellor is also a significant part of self-awareness.

More specific to sexual minorities, homophobic attitudes among counsellors (straight and gay) can evoke negative feelings within themselves towards their sexual minority clients. Practitioners may have a moral opinion against homosexuality. When this is overtly stated to the client, then it is part of discrimination and harmful practice with gay clients. However, often counsellors may not state it openly, but it may exist covertly and may keep coming up as negative judgements, inability to build rapport or other forms of resistance to gay clients. Counsellors may experience shock or repulsion when sexual acts are described by their gay clients. On the other hand, there may be excessive curiosity about the client's life, especially sexual life because it is so different from one's own. All these would hamper a genuine counselling relationship and hinder the process of counselling/therapy.

Erotic transference towards same-sex counsellors, i.e. client/s having feelings of sexual attraction towards the counsellor has been seen to be a common issue. If the counselling space has been one of the first places where the client has felt accepted and has received a lot of positive regard, then often feelings of gratitude and fulfilment evoked within the clients tend to slip into romantic attraction for the counsellor. This may be difficult for counsellors to deal with, especially if the counsellor has not worked through their own homophobia. While the same is possible between the client and counsellor of the opposite sex, the fact that it is a heterosexual attraction may make it easier for the counsellors to address it simply because these kinds of issues are more likely to be talked about and dealt with in the counsellor training or supervision group. Whereas in the case of erotic transference of a gay client, not having had the opportunity to learn about it during training or supervision would pose a challenge. Also, not having the same sexual orientation as the client may act as a stumbling block for straight counsellors. Counsellors who have not dealt with their own homophobia may experience extreme discomfort and even disgust on knowing about their client's attraction for them. Therefore, GACP advocates that heterosexual practitioners must be aware of the possibility of same-sex attractions, explore their feelings, particularly discomfort about it within supervision. Self-awareness about one's own attitudes, feelings and thoughts; positive and negative evoked by clients as well as awareness of clients' feelings and emotions towards oneself is an important aspect of good practice according to GACP. Being aware that bias and prejudice can colour counselling interactions is important as well as realizing that, in a counselling relationship two or more 'real' people with 'real' feelings are involved and hence transference and counter-transferential feelings are inevitable. All counsellors are therefore encouraged to seek supervision and create this space for reflection and self-growth.

Another related question that often tends to plague counsellors while working with all clients, but specifically in context of clients from marginalized groups is that of 'self-disclosure'. Is it necessary as a counsellor for me to disclose my sexuality to LG clients? Do I have to disclose my relationship status with them? The answer to this is really multilayered. First of all, how does the counsellor view the idea or as some may say even therapeutic technique of self-disclosure in their general practice? There are MHPs, who are trained to view themselves as a blank slate on which their clients can project their inner material, which is then processed in therapy. A counsellor trained in this tradition may avoid self-disclosure about any aspect of their lives completely. Then there are other practitioners, who for instance are trained and view the world from a feminist lens, where self-disclosure by the counsellor is seen as an important therapeutic tool. It comes from the value and belief in the feminist slogan of 'personal is political' as well as the need to equalize power relationship between the therapist and the client (Brown and Walker 1990). If a counsellor works from this location then likely that they would use self-disclosure as a tool of minimizing therapist power in psychotherapy. On the whole, depending on their training background and context of work, the degree of self-disclosure may vary and yet most practitioners may use some degree of self-disclosure rooted in the idea of 'genuine engagement' with their clients. Also purpose and timing of disclosure is an idea that has been considered in psychotherapy literature (Knox and Hill 2003). The purpose of self-disclosure is never that of the counsellor's own need but instead about the need of the client and appropriateness of the same for the therapeutic relationship.

While working with LG clients too, counsellors could keep some of the above discussions in mind to decide how much, when and why would they use self-disclosure. In the context of sexuality, another aspect that becomes significant is that of counsellors' own levels of comfort with their own sexuality. There may be counsellors who are not comfortable taking on labels such as 'heterosexual', 'straight', 'queer', 'lesbian', 'gay' and so on or may not be comfortable discussing any aspect of their sexuality even if it may seem relevant for the therapeutic work in the session. For instance, there may be a client struggling with internalized homophobia and isolation and the gay counsellor talking about their own journey towards self-acceptance maybe relevant and beneficial for the client; it may reduce isolation of the client, give them hope and a role model and so on. However, the counsellor may not be ready yet to share their journey. The same may apply to a straight counsellor working with a young woman struggling with some aspect of her sexuality, for instance an experience of abortion or premarital sex and associated guilt. These aspects related to counsellor self-disclosure and comfort about the same or deciding on whether, when, how much to disclose are some of the issues that can be resolved in ongoing supervision.

In conclusion, ethical practice with lesbian and gay clients certainly starts with refusing to engage in any form of treatment focused on changing or undermining same-sex sexual orientation in any manner. It extends to much more in terms of educating oneself on sexual minority concerns, resources for LG clients and working on own biases and prejudices when it comes to issues of non-normative sexual expressions.

References

American Psychiatric Association. (1952). *Diagnostic and statistical manual of mental disorders*. Washington DC: APA.

American Psychiatric Association. (1987). *Diagnostic and statistical manual of mental disorders (3rd edition revised)*. Washington DC: APA.

American Psychological Association. (2000). *Guidelines for psychotherapy with lesbian, gay, and bisexual clients*. APA.

Brown, L., & Walker, L. (1990) Feminist therapy perspectives on self disclosure. In G. Stricker & M. Fisher (Eds.), *Self-disclosure in the therapeutic relationship*. New York: Plenum Press.

Davison, G. C. (2001). Conceptual and ethical issues in therapy for the psychological problems of gay men, lesbians, and bisexuals. *Psychotherapy in Practice, 57*(5), 695–704.

Fernandez, B., & Gomathy, N. B. (2003). *The nature of violence faced by lesbian women in India*. Mumbai: Research Centre on Violence Against Women, Tata Institute of Social Sciences Mumbai

Ghosh, S., Bandyopadhyay, B. S., & Biswas, R. (2011). *Vio-map: Documenting and mapping violence and rights violation taking place in the lives of sexually marginalized women to chart out effective advocacy strategies*. Kolkata: SAPPHO for Equality.

Haldeman, D. C. (1994). The practice and ethics of sexual orientation conversion therapy. *Journal of Consulting and Clinical Psychology, 62*(2), 221.

Hooker, E. (1957). The adjustment of the male overt homosexual. *Journal of Projective Techniques, 21*(1), 18–31.

Kalra, G. (2012). Pathologising alternate sexuality: Shifting psychiatric practices and a need for ethical norms and reforms. *Indian Journal of Medical Ethics, 9*(4), 291.

Kalra, G., Gupta, S., & Bhugra, D. (2010). Sexual variation in India: A view from the west. *Indian Journal of Psychiatry, 52*(7), 264.

Kinsey, A. C. (Ed.). (1953). *Sexual behavior in the human female*. Indiana University Press.

Kinsey, A. C., Pomeroy, W. B., & Martin, C. E. (1948). *Sexual behavior in the human male*. Indiana University Press.

Knox, S., & Hill, C. E. (2003). Therapist self-disclosure: Research-based suggestions for practitioners. *Journal of Clinical Psychology, 59*(5), 529–539.

LABIA. (2013). *Breaking the Binary: Understanding concerns and realities of queer persons assigned gender female at birth across a spectrum of lived gender identities*. Mumbai: LABIA. Retrieved March 10, 2016, from www.labiacollective.org

Mehta, M., & Deshpande, S. N. (1983). Homosexuality—A study of treatment and outcome. *Indian Journal of Psychiatry, 25*(3), 235.

Meyer, I. H. (1995). Minority stress and mental health in gay men. *Journal of Health and Social Behavior, 36*(1), 38–56.

Meyer, I. H. (2007). Prejudice and discrimination as social stressors. In I. H. Meyer & M. E. Northridge (Eds.), *The health of sexual minorities: Public health perspectives on lesbian, gay, bisexual and transgender populations* (pp. 242–267). Springer Science + Business Media, LLC.

Parekh, S. (2003). Homosexuality in India: The light at the end of the tunnel. *Journal of Gay & Lesbian Psychotherapy, 7*(1/2), 153.

Pradhan, P. V., Ayyar, K. S., & Bagadia, V. N. (1982). Homosexuality: Treatment by behaviour modification. *Indian Journal of Psychiatry, 24*(1), 80.

Ranade, K. (2015a). Medical response to male same-sex sexuality in Western India: An exploration of 'Conversion Treatments' for homosexuality. In A. Narrain & V. Chandran (Eds.), *Nothing to fix: Medicalisation of sexual orientation and gender identities*. Delhi: Sage Yoda Press.

Ranade, K. (2015b). 'Growing Up Gay' In a Heterosexually Constructed World: Developmental Life Histories of Young Gay and Lesbian Persons in Urban India, (Unpublished Doctoral Dissertation). Tata Institute of Social Sciences-Mumbai. Retrieved March 10, 2016, from http://shodhganga.inflibnet.ac.in/handle/10603/55840?mode=full

Rao, T. S., & Jacob, K. S. (2012). Homosexuality and India. *Indian Journal of Psychiatry, 54*, 1–3.

Ritter, K., & Terndrup, A. I. (2002). *Handbook of affirmative psychotherapy with lesbians and gay men*. New York: Guilford Press.

Sakthivel, L., Rangaswami, K., & Jayaraman, T. (1979). Treatment of homosexuality by anticipatory avoidance conditioning technique. *Indian Journal of Psychiatry, 21*(2), 146.

Shaw, E., Butler, C. A., Langdridge, D., Gibson, S., Barker, M., Lenihan, P. & Richards, C. (2012). *Guidelines and literature review for psychologists working therapeutically with sexual and gender minority clients*. British Psychological Society.

Silverstein, C. (1991). Psychological and Medical Treatments of Homosexuality. In J. C. Gonsiorek & J. D. Weinrich (Eds.), *Homosexuality: Research implications for public policy* (pp. 101–114). Newbury Park, CA: Sage Publications.

Silverstein, C. (1996). History of treatment. In R. P. Cabaj & T. S. Stein (Eds.), *Textbook of homosexuality and mental health* (pp. 3–16). Washington, DC: American Psychiatric Press.

Smart, L., & Wegner, D. M. (2000). The hidden costs of stigma. In T. F. Heatherton, R. E. Kleck, M. R. Hebl, & J. G. Hull (Eds.), *The social psychology of stigma* (pp. 220–242). New York: Guilford Press.

Smith, G., Bartlett, A., & King, M. (2004). Treatments of homosexuality in Britain since the 1950s —an oral history: The experience of patients. *British Medical Journal, 328*(7437), 427.

Verghese, A. (2014). A fresh look at homosexuality [Letter to the editor]. *Indian Journal of Psychiatry, 56*, 209–210.

World Health Organization (1992). *ICD-10 classification of mental and behavioural disorders*. Geneva: WHO.

Yogyakarta Principles (2007). *Principles on the application of international human rights law in relation to sexual orientation and gender identity*. Retreived March 10, 2016, from http://www.yogyakartaprinciples.org/

Chapter 10
Employee Assistance Programmes: Emergent Ethical Issues

Ellen Mrinalini Shinde

Abstract Organizations in India and across the globe are adopting employee assistance programmes (EAP) to provide confidential counselling and consultation services to address personal problems experienced by employees and/or their family members. EAP providers and counsellors have a dual responsibility towards meeting the needs of clients in counselling and the needs of the company which employs the EAP service. Balancing the goals of 'helping' with the need to focus on business profit and 'return on investment' poses multiple challenges. The primary ethical dilemma centres on identifying whose needs take primacy. Some of the salient concerns relate to: Protecting the confidentiality of the individual client versus the risk of lost business; Compromised quality versus larger profit margins (for the EAP provider); Confidentiality of the employee (client) versus risks and safety of the company (client); Quality of the EAP service versus securing new business contracts (for the EAP provider). The EAP counsellor has to grapple with conflicting goals, and ensure safety and therapeutic value for the individual client even if it comes at an additional cost to the EAP provider or the company. EAP counsellors need high levels of competence not only in creating therapeutic outcomes but also in understanding the world of work. Well thought through ethical processes should be part of the foundation of the counselling practice in the EAP and guide the business plans of the EAP provider.

Keywords Ethics · Employee assistance programmes · Workplace mental health

E.M. Shinde (✉)
1to1help.net Pvt Ltd, Bengaluru 560038, India
e-mail: ellen@1to1help.net

P. Bhola and A. Raguram (eds.), *Ethical Issues in Counselling and Psychotherapy Practice*, DOI 10.1007/978-981-10-1808-4_10

10.1 Employee Assistance Programmes: A Growing Trend

A recent World Health Organization (2005) report called for the development of policies and strategies to promote the mental health of employees and prevent and treat mental health problems. The links between mental health issues and reduced productivity and increased costs make a strong business case for workplace mental health policies. There is a growing recognition that work can contribute to personal wellbeing, and equally, positive mental health among employees can impact various indices of workplace functioning and growth.

Organizations in India and across the globe are adopting employee assistance programmes (EAP) to provide confidential counselling and consultation services to address personal problems experienced by employees and/or their family members. Jindal (2014) discussed the growing trend for EAP services among large multinational companies in India, covering a range of industries. In addition to the beneficial impact on workplace productivity and employee attrition, the provision of EAP services can enhance the perception of the organization as a caring workplace and also serve to protect from litigation (Buon and Taylor 2007).

The organization either outsources these preventive and intervention services to external providers or sets up an in-house service. EAP services typically cover domains including job stress, relationship issues, substance abuse and a range of mental health concerns. These services are typically offered free of cost to employees and the duration of these services are often limited. Buon and Taylor (2007) surveyed employee perceptions of employee assistance programmes in the United Kingdom and Europe. The majority of respondents felt that a wide range of services should be offered: telephone counselling for employees, face-to-face counselling for employees, telephone counselling for employees' family members, alcohol and drug support and/or counselling, critical incident and trauma counselling, stress management and work-life balance services.

The employees may directly contact the EAP service or may be referred by a supervisor or manager. The employee could approach the EAP team through the helpline number or other online avenues. In certain situations, the supervisor may find a decline in the employees' productivity and may recommend taking EAP assistance. The supervisors are trained to identify issues that are appropriate for referral and ways of sensitively approaching the employee. In case of a supervisor referral, information about employee participation or non-participation may be shared, but the contents of the sessions are private and confidential.

Several EAPs do not employ full-time counsellors to serve their client base. Instead, there are counsellors appointed by the EAP to serve on an affiliate counsellor panel. The counsellors are free to pursue other full-time/part-time practice options. Some EAPs, however, employ both full-time staff and also affiliate counsellors. The affiliates may not be involved in all of the EAP services. The full-time counsellors typically do more than just the counselling aspects of EAP. Their work includes creating online resources, conducting workshops and training seminars with groups of employees. EAP counsellors also need to ensure that response protocols for critical clients are in place when there is a risk of suicidal or homicidal behaviour or abuse in significant relationships. The adequacy of protection and support for the employee and reduction of the risk of liability for the client company also fall within the scope of their work.

The actual working of the EAP with individual affiliates follows standard procedures. This ensures that the client is served well. The EAP assigns clients to the affiliate counsellors allowing them to schedule in sessions depending on the client and counsellor's mutual convenience. There are usually standard operating procedures and formal guidelines that govern the counsellor-client interaction under the EAP covering, as well as the reporting and documentation expectations. These standard operating procedures allow EAPs' to meet client requirements (as per agreements signed) of both the companies and the individual employees.

Remuneration to affiliates is paid on a per session basis. EAPs typically make these payments at the end of a month following the submission of the session notes/summaries by the affiliate to the EAP. EAPs usually appoint an Affiliate Counsellor Manager to help coordinate all the activities of the affiliate counsellors. These include identification of the number and locations of the counsellors needed to serve the EAP, the selection and recruitment of affiliate counsellors (based on predetermined eligibility criteria) the training of the affiliates, evaluation of work done, retraining and additional skills training, and providing clinical supervision.

10.2 EAP Services: Ethical Dilemmas and Decision-Making

EAP services represent an emerging professional playing field for mental health practitioners, accompanied by unique ethical and practice issues (Schonberg and Lee 1996). The unique "dual client" nature of employee assistance work can complicate ethical decision-making because the EA professional always has simultaneous client responsibilities to both the work organization and the individual employee or family member. The counsellor/therapist often has to manage multiple roles; providing psychological services to the employees as well planning other professional services to the company such as conflict management or prevention or training of managers. There are multiple stakeholders in professional situations; the organization providing EAP services, the EAP counsellor/therapist, the employee seeking help, and the employer organization. Balancing the ethical obligations to each player in the equation and integrating the business needs with the professional code of ethics can be complex and challenging.

While mental health practitioners; counsellors, clinical psychologists, psychiatric social workers and psychiatrists, are guided by professional ethics codes, the EAP context may give rise to unique ethical dilemmas concerning confidentiality and access to client records, consent, conflicts of roles or interests and remuneration for services. This new arena of work could give rise to uncertainties about how to translate professional ethical guidelines into practice.

The ethical challenges are not restricted to the counselling sessions with employees but begin at the outset of the business relationship between the EAP provider and the company. While employee wellbeing is an important contributor to workplace productivity, the EAP team must be careful not to oversell the possible impact of their inputs.

The chapter illustrates dilemmas based on experiences working in the field and clarifies the ethical positions and decision-making pathways.

Case 10.1
You are an EAP affiliate provider. You are seeing a self-referred client, who reports that he is in a clandestine sexual relationship with the company president, and has been offered a promotion by the president. He does not believe he is qualified for the position, but plans to accept it. He is concerned about the reaction of his better qualified peers. He has been having anxiety about this issue, and is coming to you for strategies to manage his anxiety. Your contact with the company, the Human Resource (HR) Director, has become a good friend, and often calls you for advice. Shortly after your first session with the client, she calls and confides that she suspects the president is sexually harassing employees and asks you for guidance on how to deal with the situation.

The first step is to clarify your position vis-à-vis the company and the employee/client (see Case 10.1). An EAP counsellor provider is a part-time worker and someone who is not bound by contract to the company. The counsellor is only part of the counselling process and is not involved with managerial processes.

The ethical dilemma here is that of confidentiality and appropriate disclosure of the information received from your client and from the HR Director. As in most ethical dilemmas, the decision-making must consider the individual merits of the case. The therapist may encourage the employee to reflect on the future implications and consequences of his decision and the possible difficulties arising out of a personal relationship across the power hierarchy in the professional context. While dealing with the revelations by the HR Director, the therapist may suggest increasing awareness about sexual harassment at work place and the provision of appropriate processes and channels for reporting.

Although there are limits to confidentiality between therapist and client, this employee revelation needs to be protected by the therapist, even in the face of implicit or explicit pressures from the employer. Any confidentiality dilemmas must of course be viewed in the context of any laws that clearly outline the obligations for the professional care provider.

Case 10.2
You are the part of the management team for an organization that provides employee assistance services to companies. You have been invited to submit a proposal to a highly desirable client company. You have been highly recommended by a close confidante of the company president, and have every reason to believe you will receive the contract, provided that you can meet the requirements. The contract will be highly lucrative for you, at a difficult time financially. The company is insisting that you add their previous panel of providers, who are not well versed in EAP and not well known to you, to your affiliate provider panel. Historically, this has been against your company policy. What is the best, and most ethical, way forward?

The ethical dilemma for the EAP provider represents the pulls and tugs between the business agenda and the obligation to ensure the competence of services provided to distressed employees (see Case 10.2).

The EAP provider has a strict policy on the pre-identified qualifications required to select counsellors as part of the affiliate provider panel. These include both academic degrees and the content and duration of work experience. For example, counsellors who primarily work with school children may not be selected as the EAP work involves interface with adults. Should the EAP organization compromise on the quality checks and processes for their counsellors? An ethical EAP organization would adhere to their standard procedures to ensure quality services, even though this might pose a risk to the business bottom line.

Case 10.3
As founder of a growing organization, you have a contract to provide EAP services to a large company. The company has recently hired a new management team, which has made many changes in the company. Subsequently, there has been a dramatic increase in formal management referrals. Your counsellors have reported that many of the referrals do not appear to have apparent performance or behavioural issues. These employees have one thing in common—they have all been openly critical of the changes in the company. They report being previously warned by supervisors that complaints about the company processes would result in an EAP referral. What do you do?

Most manager referrals appeared to be disciplinary consequences for 'problematic employees' who had openly aired their grievances about the new management (see Case 10.3). The external EAP provider would need to walk the ethical tightrope in addressing this situation. Staying true to their 'counsellor ethics' would mean not letting the fact that they were retained (and paid) by the management to play a role in how they dealt with employees during sessions.

The EAP provider might also consider reiterating the process and parameters of manager-led referrals within the organization, without referring to the in-session revelations. While this scenario highlights one type of inappropriate referrals, there are other scenarios as well. Harley (1991) examined biases in the referral process and noted a tendency to make referrals following triggering incidents at the work site, using EAP as a 'last resort' without addressing possible organizational or external factors. Supervisors' moral judgments may also affect their ability to decide who is an appropriate candidate for an EAP referral (Besenhofer and Gerstein 1992). To facilitate appropriate referrals, there is usually a panel to decide on referrals, consisting of the HR and managerial staff. Disciplinary actions do not fall within the gambit of EAP work and this ethical line must be adhered to. If the management brings up issues related to disgruntled employees, the EAP provider could recommend methods to address this at a systemic level.

Case 10.4

You are an EAP counsellor and a self-referred employee comes to see you. He works in a chemical plant, operating gauges that are critical to the safe control of very volatile products. The employee tells you that he is worried about his cocaine use, which he initially used intranasal but now uses intravenously on a daily basis. He reports that he uses only during non-work hours, but lately has a hard time making it through a shift without becoming listless and preoccupied with craving his next use. His speech is rapid and his behaviour agitated as he meets with you.

Your assessment leads you to conclude that he will likely require residential treatment to recover from his addiction. When you present your assessment and recommendation to him, he becomes upset and says, "Heck no, I thought I could just talk to someone once a week". He gets up disgruntled, states that he is late for his shift, and walks out.

The company has created no policies that address this scenario. You know that there are codes of conduct and mental health statutes that pertain to a client expressing imminent homicidal or suicidal threat. But there is no imminent threat here; it is only your awareness that your client may, due to drug-induced impairment, pose an unintentional threat to the safety of co-workers and the general public. What do you do?

The limits of confidentiality are clearly spelt out in most professional ethical codes and pertain to an evaluation of risk to self or others, usually in the contexts of suicidal or homicidal intent (see Case 10.4). Other revelations from clients, for example, about a falsified resume, would not be disclosed to the company. As a counsellor who feels that the drug abuse could pose a threat, however ill-defined, for colleagues in the workplace, do you go ahead and inform the company? Counsellors must guard against a knee-jerk response and consider various options.

One possible solution would be to try and empower the individual in counselling to request his supervisor for personal or medical leave (without revealing the details). In the event that the client fails to meet the specified deadline with his supervisor, the therapist could choose to inform the client of the need for disclosure to the supervisor because of the potential danger to self, others and the possible liability issues for the company. Another option would be to disclose the information to the designated emergency point of contact within the company. Indiscriminate and immediate disclosure to the company may not benefit the company (possible loss of an employee), the client (possible loss of job) or the therapist (possible compromise of professional ethics). Another possible path is to discuss and disclose the information to the employee's family member. Each of these decisions need careful consideration and must be based on a combination of factors; employee preference, and the costs, benefits and negative consequences of each of the chosen paths.

Case 10.5

The General Manager (GM) of the company tells you that you can have the EAP contract, provided that you will reveal the names of the people who have used the service. The names are to be put in an envelope and released to the GM.

You are about to make inroads into a new kind of industry. All you have to do to get the contract is give the names to the GM ...or lose the contract. What should you do?

Despite business pressures, the ethical commitment to clients is paramount and cannot be compromised (see Case 10.5). Companies may lay down conditions and EAP providers might struggle against profits becoming a major determining factor in their contractual agreements.

An organization shared its response to this real-life scenario; The General Manager was clearly told that the names of clients using EAP services could not be disclosed. The ethical framework was discussed with the organization and there was a positive end to this story. The EAP provider got the contract on the basis of their professional integrity. Further, the competence and quality of their services resulted in renewed agreements with the company.

In a related scenario, senior management may enquire if a particular employee, noticed to be causing problems at work, has registered for EAP. Such issues may be resolved by reiterating policies about non-disclosure. The senior management may be advised to consider a formal referral, if warranted, with clear reasons for referral. Subsequently information about attendance at sessions could be shared.

Employers are entitled to access information about the number of employees who have used the service, the types of problems dealt with, the duration of treatment and the rationale for fees billed to the company. In some instances, the size of the workplace may mean that employees who access EAP are more easily identifiable and the EAP provider must define procedures to ensure professional secrecy (Ordre des psychologues du Québec 2005). A third-party audit process could check on the usage statistics without compromising on client confidentiality.

Case 10.6
A client comes to you and talks about a senior colleague involved in a fraud related to awarding vendor contracts. When the senior colleague was caught, he managed to conceal many details and passed it off on employees who had recently joined the organization and were therefore 'unaware of the procedures'. Your client is one of these employees and is concerned about how to deal with this situation as well as anything that might come up in the future. The situation is complicated by the fact that the senior colleague is in your client's direct line of reporting. How can you help him?

The EAP counsellor may often need to engage with the client to discuss delicate workplace issues that may involve persons in senior management positions (see Case 10.6). In this scenario, the counsellor and employee may discuss a range of issues: the options for disclosure and non-disclosure, whom to disclose to, possible outcomes and risk to the client, and the possible emotional and other impacts on the client.

The EAP provider may present a camouflaged 'hypothetical scenario' to company personnel, without sharing the confidential information divulged by the client. This would enable exploration of the most effective pathways for disclosure, with minimal risk to the client. This information could be fed back to the client to help him in the eventual decision regarding disclosure. The essential message is that confidentiality is due to the employee, even though the company is also a client.

Case 10.7
The company that the client works in supports only six sessions with an EAP therapist. However, your client is at a crucial stage of progress in his therapy and today is his final session. You know that your work together is incomplete… what do you do?

In this scenario, there are ethical obligations on two fronts; towards the company and towards the employee (see Case 10.7). The EAP provider should not contravene the contractual arrangement with the company and provide additional sessions. If the client needs to extend the counselling sessions, the EAP provider could consider seeking permission from the company but this is usually recommended in select situations; for example if the client is suicidal or when there has been a death of a family member. The EAP could provide information about paid counselling services that may be available and this must not be influenced by financial or business considerations. Some organizations that offer EAP services ask their external affiliates to sign agreements that they would not continue with EAP referred clients in their private practice if ongoing sessions are needed. Instead, they would refer them to the EAP provider who would provide a range of options to the client who would take an autonomous decision (see Case 10.8).

Case 10.8
The senior Vice President walks into your office. At first you think this is related to some administrative work but it turns out he is seeking your professional help as a therapist. You are concerned whether the work that is going to happen over the next couple of sessions with him is going to influence your evaluation. Do you make more of an effort, providing longer or more frequent sessions or going out of your way, when it comes to this client?

From an ethical standpoint, the therapist must deliver the same quality of service to every client. This issue needs to be stressed while training affiliate providers; every employee must receive the same professional care, regardless of position in the organization, age, gender, sexual orientation, religion or other factors. Although the company and EAP provider are in a business relationship, one which is subject to time-bound reviews and renewals, the client's power position in the company should not cast an undue influence.

Case 10.9
As founder of a well-established EAP organization, you have a strong and ongoing business relationship with a global company. In recent discussions, the management has requested you to be on the board that considers appraisals and promotions. This is part of their initiative to consider the fit between employee personality and the skill sets required in specific job roles. This is a prestigious offer and you wonder if you should accept it.

The EAP counsellors must be cognizant of potential conflicts of interest arising from dual or multiple relationships (see Case 10.9). The credibility of the EAP services could be impacted if employees wonder if the therapeutic services offered are indeed safe and confidential. Employees who have availed of EAP services would have concerns about the influence that this might have on their performance evaluation. The ethical guidelines recommend that such dual relationships should be avoided by EAP service providers.

Case 10.10
An employee reports difficulties with sleep and appetite, reduced energy levels and feeling 'stressed all the time'. He dates this to the introduction of rotating shifts in the organization about two months ago. As a counsellor you also feel that a large part of his difficulties may be attributable to the disruptions in circadian rhythms and wonder of you should share this information with the management. What should you do?

As an EAP service provider, the primary role is to address the employee's concern on an individual basis (see Case 10.10). This does not imply that problems at the organizational level are never to be discussed. A more measured approach would be to report repeated trends seen across sessions with employees. For instance, of rotating shifts seem to be having significant negative consequences on the physical and mental wellbeing of 15 clients from the company, this issue may be conveyed to the management, without revealing individual identities.

As we see from the case vignettes, the EAP counsellor must be trained not only to be competent in the interventions necessary but also in understanding the work contexts of the clients in counselling. It is not sufficient for the counsellor to only understand the concerns of the individual and their impact. He/she must be able to understand the company's investment in EAP and its expected return in terms of productivity. However, at no point, can the confidentiality of the individual client be compromised. The counsellor strives to provide the best both to the individual client and the company which also is a client. In doing so, the EAP counsellor must be very careful in understanding the problems/concerns presented in counselling to meet the needs of the individual employee as well as protect the company by doing everything possible to minimize liability risks whenever it is necessary. The EAP counsellor must clearly understand where confidentiality of the individual needs to be protected, and when the company needs to be served in case of potential liability.

10.3 Ethical Questions for EAP Service Providers

The ethical decision-making process can be aided by unambiguous answers to the following questions:

1. Who is the client? The individual or the organization?
2. What defines decisions? Profitability or client wellbeing?
3. Who services the client?
4. What determines the mode of counselling? Client concerns and preference or business need?

There is a contractual agreement between the EAP provider and the company but the primary client is always the employee. The needs of the client must not be subverted by concerns about profitability. The competence of the counsellors is an important ethical question and organizations providing EAP services need to have clear cut guidelines for their selection. A base qualification for selection of counsellors is essential and this is something that clients must be aware of. Counsellors are required to sign a confidentiality agreement along with their appointment letter.

Operating procedures need to include standardized procedures for uniform training of affiliates with continual monitoring and supervision of their sessions. Ongoing staff training can ensure competence and growth and development among counsellors. The standardized training may target a range of skills through a stepwise process. For instance, learning to present a power-point session to employees would include the following steps; initially accompanying a senior, then reading and preparation for one's own presentation, mock sessions at the EAP office, and finally being accompanied by another experienced person to ensure the session goes as planned. This handholding can take up a lot of time but ensures quality of the staff. Manualized procedures and International Organization for Standardization (ISO) certification may be considered by organizations providing EAP services to ensure objectivity, transparency and external regulation of processes. Efforts at staff retention are also essential as this directly impacts continuity and quality of services. Measures aimed at the emotional wellbeing of EAP staff are also essential as working with clients' difficulties can be stressful for the care provider as well.

In a recent study (Csiernik and Darnell 2010), Canadian affiliate counsellors shared ethical issues in relation to inadequate training and support, apart from those related to administrative requirements like filling in forms. They also experienced difficulties in providing the mandated or required counselling inputs to clients due to pressure from their employers to spend less time with clients.

Competent EAP services would typically require all counsellors to submit session summaries, with dates of sessions. The supervisor usually scrutinizes the frequency of sessions scheduled to ensure adequate spacing for effective work to be done. The session summaries are also examined to make sure that each session creates value for the client and there is no work left incomplete, resulting in the need for additional sessions.

EAP services must define the limits of their competence and develop an adequate referral network for specific needs. Mental health emergencies like suicidal intent or psychosis may require intensive management, medication or hospitalization. Substance dependence is another area which might call for supervised detoxification and residential or rehabilitation facilities.

The modes of counselling offered can range from face-to-face sessions to telephone or internet-based services. The costs of traditional face-to-face sessions are undoubtedly higher but the decision must rest on the client's choice, comfort, and needs. Email communication can be held in the court of law so they are usually not used for sensitive issues.

10.4 Ethical Issues in Documentation and Reporting

EAP services involve large amounts of documentation and reporting and this must be done through a careful ethical lens. Processes for documentation related to counselling sessions are include the locked storage of data. All backup data must be encrypted and stored carefully. Employees in the EAP organization are usually given isolation access rights to data; for example the sales personnel may have access only to the statistical data about usage of services and not the qualitative data or any information that might identify the service user.

Usage reports are those reports that the EAP therapists have to give the company annually, semi-annually or quarterly depending on their protocols. These reports ensure confidentiality and anonymity of users. For instance, only the number of clients, age, gender (even this is sometimes neutralized) and the companies from which the clients come from are disclosed. Details such as the designation of the employees are also concealed.

EAP organizations must report usage data about their services accurately. Since this information is confidential, the company has no way to cross-check figures. It is possible that the EAP provider is tempted to exaggerate usage data, particularly if there is a significant fall in persons registering for EAP services. EAP provides its services through online articles, workshops, self-assessments and individual counselling. Utilization reports should provide a separate account on usage on each of these parameters. Accurate reporting ensures that there is a reflection on any reasons for under-utilization and concerted efforts to increase awareness and promote EAP services.

10.5 Codes of Ethics for EAP Providers

The specialized nature of work related to employee assistance services in reflected in the proliferation of ethics codes that address these issues. Codes emerging from different countries also factor in the legal frames and requirements in each country. Unfortunately, although the field is growing in India, organizations have not yet come together to formulate a code of ethics to guide their work.

Most codes suggest that the nature and modalities of the service provision be specified in writing, in the form of a contract (Ordre des psychologues du Québec 2005). This document should define the obligations of the EAP provider as well as the company to which services are delivered. While the employer may want some degree of access and control, all guidelines emphasize that any accommodations must align with the principle laid out in professional codes of ethics for mental professionals. The Ordre des psychologues du Québec guidelines (2005) describe the ethical obligations pertaining to professional secrecy, consent, conflict of roles and conflict of interest and keeping of and access to the client's record. In another document (Ordre des psychologues du Québec 2006), situational scenarios illustrating potential ethical dilemmas are debated and appropriate methods of resolution are proposed for the EAP provider. The Employee Assistance Trade Association, founded in 1986, has members from across the United States and Canada and outlines ethical practice standards with useful detailed sections on the referral process and marketing and sales of EAP offerings (Employee Assistance Trade Association 2013). Another set of guidelines (Employee Assistance Professionals Association 2010) provide useful inputs on training and professional development of affiliates and programme evaluation processes.

Every employee in the organization providing EAP services should be required to sign the code of ethics that must be adhered to. Regular training exercises using case scenarios derived from industry experience would be of value. A consultative process for discussion of ethical decision-making needs to be set firmly in place. The commitment to ethical guidelines should be seen as critical and non-negotiable by all stakeholders; the EAP provider, the company and the employee.

10.6 Conclusion

The need to serve both the individual and the company can often create conflict in the minds of young counsellors, or among those who have not been part of an EAP service. There is the need to re think goals and purposes of the counselling interventions. In serving both the individual and company adequately, one needs to grapple with the 'business of counselling'. Training of counsellors in intervention techniques, working of a company, and well-reasoned and thought through ethical principles, extensive supervision, thorough documentation, rigorous processes and audits, are all necessary to ensure quality of services meet global standards.

Research needs to focus on interventions that will ensure best therapeutic outcomes while keeping in mind the limits of counselling in an EAP context.

References

Besenhofer, R. K., & Gerstein, L. H. (1992). Referrals to Employee Assistance Programs (EMS) characteristics of hypothetical supervisors, EAPs, and substance abusing workers. *Employee Assistance Quarterly, 7*(2), 41–62.

Buon, T., & Taylor, J. (2007). A review of the Employee Assistance Programme (EAP) market in the UK and Europe. Retrieved from http://www.buon.net/files/RGUEAEF.pdf

Csiernik, R., & Darnell, K. (2010). Perceptions of affiliate EAP counselors: An exploratory study. *Journal of Workplace Behavioral Health, 25*(2), 154–168.

Employee Assistance Professionals Association. (2010). EAPA standards and professional guidelines for employee assistance programs. Retrieved from www.valueoptions.com/providers/Handbook/…/EAP_Guidelines.pdf

Employee Assistance Trade Association. (2013). Code of Ethics. Retrieved from http://www.easna.org/research-and-best-practices/what-is-eap/code-of-ethics/

Harley, D. A. (1991). Impaired job performance and worksite trigger incidents: Factors influencing supervisory EAP referrals. *Employee Assistance Quarterly, 6*(3), 51–70.

Jindal, D. (2014). Employee Assistance Program (EAP) in India. *International Research Journal of Management Science & Technology, 5*(2), 276–293.

Ordre des psychologues du Québec. (2005). "Working in the framework of Employee Assistance Programs (Part 1)." *Ethics Guidelines, 6*(5). Retrieved December 31, 2015, from https://www.ordrepsy.qc.ca/pdf/0FicheDeonto_ang_Nov05.pdf

Ordre des psychologues du Québec. (2006). "Working in the framework of Employee Assistance Programs (Part 2)." *Ethics Guidelines, 7*(1). Retrieved December 31, 2015, from https://www.ordrepsy.qc.ca/pdf/0FicheDeonto_ang_Janv06.pdf

Schonberg, S. E., & Lee, S. S. (1996). Identifying the real EAP client: Ensuing ethical dilemmas. *Ethics and Behavior, 6*(3), 203–212.

World Health Organization. (2005). *Mental health policies and programmes in the workplace*. Geneva: WHO.

Chapter 11
Intersections Between Ethics and technology: Online Client–Therapist Interactions

Kakli Gupta, Ananya Sinha and Poornima Bhola

Abstract With the growing technological advancement, the paradigms of client–therapist interaction are also changing. There is an emerging trend of practising e-therapy in India. The chapter discusses the possible ethical issues that might arise in the context of client–therapist interaction in the digital world. It enlists the steps of deciding if a client is suitable for e-therapy and how to think about the therapeutic frame in the context of e-therapy. It highlights issues related to confidentiality and electronic security, boundary management particularly with respect to time, payment and fees, contact between sessions and multitasking with e-therapy. It also highlights the ethical concerns with respect to client–therapist encounters in the online space. The ethical codes and guidelines are analysed and recommendations are made to help mental health practitioners address the ethical issues arising in this context.

Keywords Digital ethics · Telepsychology · e-therapy

The last decade has seen remarkable innovations in information and communication technology, sweeping across all walks of life. Today, it is possible to communicate with people from any part of the world at any time, both verbally as well as visually. Communication is the basis of psychotherapy and it is not surprising that the advances in communication technology have had a significant impact on the profession. While the technological advances bring a lot of flexibility to the field of

K. Gupta (✉)
Private Practice, Bengaluru, India
e-mail: kakligupta@gmail.com

A. Sinha · P. Bhola
Department of Clinical Psychology, National Institute of Mental Health and Neuro Sciences, Bengaluru 560029 India
e-mail: sinha.ananya.cs@gmail.com

P. Bhola
e-mail: poornimabhola@gmail.com

P. Bhola and A. Raguram (eds.), *Ethical Issues in Counselling and Psychotherapy Practice*, DOI 10.1007/978-981-10-1808-4_11

psychotherapy, they also pose ethical challenges which can complicate the process, effectiveness as well as the safety of the therapeutic relationship.

In India, as in many other parts of the world, there has been an upsurge of organizations as well as individual practitioners who practice online therapy. The online space also offers avenues for professional networking and development and therapists may increasingly find it difficult to lead a life completely divorced from the digital interface. Therapists may have an online presence, to different degrees, and this may extend to both professional and personal domains. Therapists may feel insulated from clients in their non-professional activities on the internet, but it is quite possible that clients also share the same online space. There is a range of ethical quandaries that might arise with respect to the therapist–client online interface and this may be deliberate or inadvertent.

Professional ethical guidelines do provide guideposts about client–therapist interactions but the online playing field is a new area for many. Technology changes frequently and there is a need to keep up with the changes and anticipate the probable ethical issues that may arise. Therapists may have questions about how to translate commonly encountered ethical issues from face-to-face therapy into the domains of online therapy and interface with clients. Professional ethical guidelines may vary in the degree of detailed information and mental health practitioners may not always be aware of the specific guidelines like the ones provided by the American Psychological Association (2013) and International Society for Mental Health Online (2009). The emergent ethical issues may be more nuanced in nature, require interpretations according the cultural context and thought beyond what is laid out in the guidelines.

This chapter will highlight some emergent ethical issues in a digitally connected world, based on experiences in psychotherapy training and practice settings and from research literature. These discussions would aim at helping counsellors and therapists to be more mindful of these issues and enhance their competence in working through them.

11.1 Terminologies in the Digital Space

Online therapeutic interactions can take many forms and the array of terminologies used in the professional literature can be daunting. Therapists and counsellors might find it useful to first familiarize themselves with some of the current terminologies used to describe online therapeutic interactions.

(1) **Telepsychology**: Telepsychology refers to the use of electronic communication technologies such as telephone, email, text messaging, videoconferencing, mobile applications and web-based programs, to provide psychological services to clients (American Psychological Association 2013). e-therapy, online therapy, internet therapy and cyber-therapy are some of other terms which are used interchangeably with telepsychology.

(2) **Synchronous and Asynchronous Communication**: Synchronous communication (for example, videoconferencing, chats, telephone) refers to those forms of media in which the two parties communicating with each other are doing so in real time. Asynchronous communication (for example, emails, online bulletin boards) refers to those forms of media in which the two parties can 'leave' messages for each other to be received and responded to at a different point of time.
(3) **Stand-alone versus Augmented use of Technology in Psychotherapy**: Sometimes, a therapist's primary and only mode of communication with their client is through technology. For instance, videoconferencing with a long distance client could be a stand-alone mode of communication. On the other hand, a therapist might augment face-to-face sessions with the use of emails or phone communication.
(4) Electronic media used to conduct e-therapy may be either **text-based;** email, chat rooms, text messaging, and listservs) or **non-text-based**; telephone, videoconferencing (Maheu et al. 2005).
(5) **Social media** refers to "the wide range of Internet based and mobile services that allow users to participate in online exchanges, contribute user-created content, or join online communities" (Dewing 2010). These typically include

   a. *Blog* which is an online journal, e.g. Wordpress, Blogger.
   b. *Wiki* which is a collective website where the content can be modified by any user using a browser, e.g. Wikipedia.
   c. *Social bookmarking* which allows users to organise and share links to websites, e.g. Reddit, StumbleUpon.
   d. *Social networking* refers to "web-based services that allow individual to construct a public or semi-public profile, articulate a list of other users with whom they share a connection, and view and traverse their list of connections and those made by others within the system" (Boyd and Ellison 2007), e.g. Facebook, Linkedin.
   e. *Status update services* refer to services that create short updates about their lives or see updates created by others, e.g. Twitter.
   f. *Virtual world content* offering game-like virtual environments, e.g. Second Life.
   g. *Media sharing sites* allow users to post videos or photographs, e.g. YouTube, Pinterest, Instagram.
   h. *Mobile applications* like WhatsApp or Telegram also provide a platform for sharing messages, photographs or videos using smartphones.

The following section describes the experiential account of the first author, Dr. Kakli Gupta. It highlights how a therapist, who began with traditional face-to-face sessions with clients and embarked on the journey of practicing online therapy. The adaptations to changing client needs, therapist life circumstances, and reflections on ethical issues and ways of addressing them will be discussed.

11.2 My Introduction to Telepsychology: A Personal Account

After completing my Doctorate in Clinical Psychology from California, I had to move back to India very suddenly for personal reasons. In California, I had spent the last two years of my doctoral training working with children using play therapy. Many of these children had a history of severe trauma and emotional disturbance. In working with them, it was helpful to have the support of one-on-one consultations with my supervisors, along with group supervision and informal peer support. When I started working with children in a non-metropolitan city in India, I felt a strong need for consultation but I could not find anyone locally available, trained in using play therapy with children, in the way I was trained in California. It is then that I contacted my supervisor in the United States (US), who generously accepted to consult with me on a weekly basis using videoconferencing. Despite the cultural differences and long distance communication, her support was extremely valuable to my work in India. In fact, during our videoconferencing sessions, I would often forget that we were talking across thousands of miles. I attribute this not only to the fact that she and I had worked face to face for two years in prior to the online communication, but also to the fact that we maintained the supervision frame just as we would have in face-to-face sessions. We started and ended on time, we met (online) once a week at the same time and minimized the tendency to interact in between sessions through email. Gradually, my practice grew and with the support of my consultant, I was able to extend my work to adults as well.

Two years later, a move to another city, Bengaluru, posed a challenge in terms of providing continued service to the clients I had been seeing for a while. Some of these clients had a strong therapeutic alliance with me and local referral resources could not be identified. Since my own experience with my consultant over video-conferencing had turned out to be very positive, I extended the same option to some of the adult clients who I felt could handle the change in modality and who did not present with life-threatening symptoms. Many of the clients took up this offer and were very relieved to not have to shift to another therapist. Unfortunately, I had to terminate with most child clients, except for one thirteen-year-old girl, who was fairly verbally expressive and wanted to continue with me over videoconferencing. It was heartening that two of the other child clients who I had to terminate with, called me seven years later, as teenagers, requesting sessions through videoconferencing.

After about a year of being in Bengaluru, twenty percent of my practice was through videoconferencing with clients outside Bengaluru and eighty percent was face to face with local residents. Meanwhile, I continued working with my consultant once a week through videoconferencing. Parallel to the supervision process, my videoconferencing sessions with the clients maintained the ‘therapeutic frame’ strictly; sessions started and ended on time, the fee structure was maintained and clients paid through online banking, the physical setting of the room was consistent and confidential and online written communication was used only for scheduling and payment. In Bengaluru itself, there were times when clients would have to move to a different city or country or travel for extended periods of time. Once again, the option of videoconferencing allowed continuity in therapeutic work.

In the last few years, I have got many referrals for clients of Indian origin, working or residing in another country, who prefer to speak to a therapist from their own culture, and/or to a therapist who has lived in India as well as another country and understands the psychological struggles of immigration. In most of these cases, the client has met me face to face in Bengaluru for a couple of sessions or more before switching to videoconferencing. I prefer to see them face to face at least once to assess whether they are good candidates for therapy through videoconferencing or not.

Lately, I have had two referrals from clients who live in the city, but are not able to come to my office regularly due to physical disability or a medical condition which does not allow them to travel easily. I have met each of these clients face to face for the first session and have continued to work with them through videoconferencing since then.

As the city of Bengaluru faces serious difficulties with traffic, the commute time poses a barrier to some clients who find it difficult to keep up with the frequency of recommended sessions. In such situations, the arrangement of alternating sessions between the face to face and videoconferencing modality, works as a practically viable alternative. For other clients who absolutely need the physical presence of the therapist and the safety of the therapy room, face-to-face sessions are the only option.

As the last few years have turned out, about forty percent of my practice is through videoconferencing with clients in and out of Bengaluru, while sixty percent of my practice is through face-to-face sessions. In the last nine years of working in India, I have also been able to attend a few specialized continuing education courses in the US using videoconferencing technology. I continue to take support from consultants in the US for my own practice based on their specializations. As a professional, it makes me feel supported and equipped. I know I can reach out to the right person as and when I need to, irrespective of geographical boundaries.

Of course, using technology has not been without its challenges. There are many ethical and clinical issues that have arisen for me during the last few years. While there are definitely some generic guideposts, the individuality and unique situation of each client also come into play.

The following section draws from experiences in training and practice contexts and from research literature on ethical issues in telepsychology and discusses possible guidelines and recommendations for practitioners.

11.3 Online Client–Therapist Interface: Recognizing and Responding to Ethical Dilemmas

11.3.1 Is e-Therapy an Appropriate Choice for Everyone?

As a therapist, I am often confronted with a dilemma, both clinical and ethical, when there is a referral from a new client for ‘long distance’ therapy, for something like 'depression', or 'relationship problem'. The client may be from a geographically distant location, perhaps with limited availability of mental health professionals in that area. If one agrees to take the client on without having had a chance for a face-to-face assessment, there may be possible risk factors for which long distance therapy might be contraindicated. Conversely, refusal to take on the client may be accompanied by the guilt of not helping someone who reached out and has few options available locally.

The overarching ethical principles of beneficence and non-maleficence must be considered while assessing client suitability for any intervention, and this includes the e-therapy modality. In my practice, I almost always agree to work with a client using videoconferencing only after having had an assessment session (preferably face to face) to gauge whether or not the client would benefit from long-distance therapy. Over the years, I have found it useful to educate a potential client that I would need to make an assessment to judge whether or not 'videoconferencing' would be beneficial as a primary mode of therapy (Mallen et al. 2005).

The assessment for suitability for e-therapy is based on the following four criteria: (1) the nature and severity of presenting concerns of the client (2) the availability and comfort with using technology (3) availability of a private and confidential space in the client's home/work space and (4) the verbal expressiveness of the client, allowing for a good enough online communication.

The nature and severity of presenting concerns of the client: Are all presenting concerns amenable to treatment through online modalities? Possibly not. One would guess that both the nature as well as the severity of the problem would determine the suitability of a client for e-therapy. Typically clients with psychotic disorders, severe personality disorders and those who are suicidal and homicidal, are not considered to be good candidates for e-therapy (Ragusea and VandeCreek 2003). In addition, victims of sexual abuse and intimate partner violence may not be most appropriate for e-therapy (Bloom 1997). Suler (2001) writes,

> Tendencies towards poor reality testing and strong transference reactions may become exacerbated in text communication, thereby making them difficult to manage and potentially destructive to the treatment. People with borderline personality disorders often challenge the boundaries of therapy, which can be problematic in email communication and when combining different methods of communication (p. 678).

It is also true that any client who has been screened and found suitable for e-therapy, could manifest with suicidal intent or plans at a later stage during the course of therapy. A therapist has to always keep that possibility in mind and be prepared with contact information available for family and identify professionals who could be locally available to the client in such emergency situations (Manhal-Baugus 2001).

Availability and comfort with using technology: For online therapeutic interactions to work effectively, it is important to ensure that the client has adequate knowledge of computer functioning, has easy internet access and adequate internet speed available. These may pose a challenge in some locations in India where both computer literacy and access might be limited. The use of public 'cyber-cafes' for client–therapist online interface is fraught with ethical concerns regarding confidentiality. Usually, younger clients (12-50 year old) are more technologically savvy than older ones (50 years and above). As a therapist, one may need to keep oneself updated regarding the advances in videoconferencing modalities, and be willing to try a medium that his/her clients find familiar, useful and stable.

Availability of private and confidential space in client's home/work space: During e-therapy, the therapist may have little control over the physical environment of the therapeutic space and in ensuring a physically safe and confidential space. As a therapist, I understood the depth of this only when a client who worked with me face-to-face for two years had to shift to videoconferencing because both of us moved out from the city we met in. During one of the videoconferencing sessions, she cried and shared that she really missed the safety of being in my office during our sessions. Now that she talked to me from her home, she was always aware of the presence of people around her whom she felt emotionally fragile around. In many homes, respecting the privacy of therapy sessions may not be a given. For instance, a parent or a sibling of a teenaged client might not see anything wrong with their being in the same room, while the client is in therapy via Skype. One of my supervisees once reported the difficulty in having Skype sessions with a woman confronted with issues of subjugation within the joint family, because the family members kept coming in and out of the room, during the session. This client could not explain to the elders in the family why she needed to lock the room for this one hour. Thus, as part of the 'assessment' for suitability of videoconferencing sessions, it might be helpful to ask the clients some very specific questions around what kind of privacy they have in their home while on the call with the therapist (Ragusea and VandeCreek 2003). One may also assess the level of assertiveness of the client around being able to ask for privacy, if that is not easily understood or given in their family.

Verbal expressiveness of the Client: Videoconferencing and other forms of e-therapy tend to be much more dependent on verbal expression than a traditional face-to-face session where non-verbal communication is also an essential component of the interactions. Videoconferencing might not be the best choice for a client who is not very verbally expressive and/or who could use more non-verbal means of expression such as art or sand play to benefit from therapy. This is also another reason why e-therapy has very limited use with young children who need to be able to use 'play' as a mode of therapeutic work. The risk of missing communication cues is much higher in text-based e-therapy using email, short messaging services or other instant messaging services (Alleman 2002; Mallen et al. 2005), and this could impact the quality of therapeutic services being offered. From an ethical standpoint, these potential limitations should be clarified to the client at the very onset at the initial phase of assessment to avoid later difficulties and hindrances in the therapy process.

11.3.2 How Do We Ensure Confidentiality and Electronic Security?

Similar to traditional face-to-face therapy, the ethical mandate of confidentiality extends to the e-therapy context but requires additional technical knowledge from practitioners. This may specifically include knowledge about the privacy settings while using a particular online forum. The use of emails, online support groups and instant messaging all leave a digital trail that can be compromised if not secured properly (Frame 1997; Kanani and Regehr 2003; Mallen et al. 2005).

Zack (2004) specifies the need to have electronic safeguards against hacking of sensitive client information stored on a personal computer or laptop. Client records can be secured using password protection systems, appropriate firewalls and document encryption. For practitioners who communicate with clients through email, the use of email encryption software is also recommended. Many practitioners may choose to use free videoconferencing software, for example Skype, and need to confirm privacy policies and third-party access regulations.

At the basic level, creating separate professional and personal user accounts is recommended if practitioners have a single laptop or desktop. Many of us synchronize our email accounts across several devices; for example, one might have a personal email account synced across a smartphone, tablet, laptop and personal computer and often be constantly logged in. Any one device that is accessible without being locked with a password could unintentionally jeopardize client confidentiality. One way to partially address this issue is to remember to log out from accounts across the various devices we use.

Electronic data is tough to delete permanently; emptying the 'recycle bin' on the laptop is not a complete solution. Zack (2004) suggests the use of 'wiping software' to permanently delete client files before disposing off old computers or laptops. If online therapy forms a large segment of one's therapeutic practice, technical support would need to be readily available.

Another caution that a therapist needs to take is in terms of educating the client about digital security. For example, the therapist sends an email or text messaging to the client whose laptop or smartphone is easily accessible to other family members. There is a possibility of any family member reading the mail by the therapist that might contain confidential information about the client. Thus the clients also needs to be aware of and practice digital safeguarding at their end. This concern of digital privacy should ideally be discussed at the very onset of e-therapy during the informed consent process (Mallen et al. 2005).

11.3.3 Do Professional Boundaries Work the Same Way in e-Therapy as in Face-to-Face Therapy?

Boundaries or therapeutic frame provide the conditions under which the work of psychotherapy can happen. Without reliable and well-defined boundaries, therapeutic work can be ineffective and even carry the risk of harming a client. However, practitioners have also long known that in therapeutic work, a therapist has to be prepared to make 'exceptions' to these boundaries if they seem helpful to the client's progress. McWilliams (2004) says,

> There is very little about therapeutic boundaries that are simple. Once the frame is clear to both parties, the security of the therapeutic couple depends on observing a mutually understood set of boundaries consistently. But, somewhat paradoxically, it is also true that the most moving and healing moments in treatment are often the times when the therapist does something exceptional, stepping out of the frame and responding to the patient with a spontaneous gesture (Winnicott 1960).

Thus, what is required from a therapist is fairly complex—set good boundaries that work in general, renegotiate these boundaries a little bit in the context of each client and when a situation arises where deviation from these boundaries is expected, the therapist has to decide (sometimes in a fraction of a second) whether or not this deviation would be helpful or harmful to the client. It is not too difficult now to imagine that working with clients using e-therapy would add another layer of complexity to this decision-making. In this next section, some of the quandaries around the frames and boundaries that may arise in the context of using video-conferencing with clients and ways to deal with them will be discussed.

Time: The session duration may be the same as for face-to-face sessions. Sometimes there might be requests for shorter, more frequent sessions, typically from clients with a lot of anxiety, who prefer to connect more frequently to feel 'contained' instead of having to wait for the whole week. Since they just have to 'log in' and not 'commute' to see the therapist, they tend to find the therapist more easily accessible (Manhal-Baugus 2001). Most times, shorter (20–30 minutes) and more frequent (twice or thrice a week) session arrangements are not helpful to the client in the long run, because (a) it can increase their dependency on the therapist (b) shorter sessions tend to be more about anxiety management rather than developing a deeper understanding. There is room for exceptions and stand-alone brief videoconferencing or phone sessions could be offered to clients in crisis situations, for example, in the context of a client's experience of physical abuse by the spouse.

Dealing with technological interruption is one of the concerns of e-therapy; these may include servers crashing, equipment malfunction or an unstable internet connection (Lee 2010). One way of dealing with it is through prior discussion about responsibility for the technical failure, if the therapist is responsible for the 'lost time' and is obligated to 'make up' for it in different ways; for instance by extending the time, rescheduling the session, or charging a prorated fee. The therapist should anticipate such occurrences and keep alternate options, like a telephonic call, available. Such anticipations and alternate options should be discussed with the client prior to initiating therapy and should be mentioned in the informed consent process (American Counseling Association 2014).

Payment and fees: Most of the clients who work with me through videoconferencing pay me at the end of the month, for all the sessions they have had with me that month. Payments are typically made through an online bank transfer, through cheque or by depositing a cheque or cash directly to my bank account. A majority of these clients choose 'the once a month, online modality option,' due to cashless convenience. I have often sensed anxiety related to payment among therapists who are new to practicing online therapy. In the last eight years of using telepsychology, I have never had an experience of not being paid for the work. There have been three to four occasions, where I have had to send reminders to clients who might have terminated with me to clear the pending dues. A few clients have insisted on paying me in advance, but that has been mostly to deal with their own ambivalence about being in therapy. They felt that they were more likely to continue to see me if they have already paid me. Overall, there is some therapeutic value in 'trusting' clients with payments and this can enhance the overall trust and safety in the therapeutic relationship.

Practitioners may wonder if the fee for therapy session should be different for a face-to-face session versus a session done through videoconferencing. I charge the same fee for a videoconference session as a face-to-face session because I feel my presence, effort and time spent with the client through either modality is the same. Charging a lower fee for videoconferencing session would imply that it is of 'less value' or 'lesser quality', which may not be the case (unless there is a significant reduction in the quality of the therapy experience due to technological issues such as lack of clarity in voice or video). In fact, there is an added cost for high-speed internet connection in addition to the usual workspace costs.

Contact between sessions: Unless a therapist has contracted to do therapy over email (in addition to or in lieu of videoconferencing/face-to-face sessions), it is best to maintain a good boundary and understanding around how between-session communication would be handled. Therapeutic 'conversations' over email should be done with extra caution, as the written word without the benefit of verbal gesture or tone of voice may be misinterpreted easily (Manhal-Baugus 2001). Sometimes it just is very hard to communicate the same amount of concern and warmth through writing that one can do in a face to face or videoconferencing or even through a telephonic session. If the client needs to contact in between sessions, it may be done through a variety of options; phone or text message or email and the policy should be discussed at the very onset of therapy. The therapist would also need to discuss about what kind of communication is acceptable (e.g. for rescheduling sessions or for sharing about distressing experiences and emotions) and clarify the type and timing of response that might be expected from the therapist.

Multitasking with e-therapy: Another issue that is unique to e-therapy is that it is technically possible for both the client and the therapist to be doing something online in parallel to listening or talking to each other. In a face-to-face session, it would be very rare for a client to be text messaging a friend while talking to the therapist. However, it is not that uncommon for therapists to find an adolescent client typing a message to a friend while also talking to the therapist on Skype. These situations can be delicate as sometimes, the therapist may have a doubt that the client is multitasking, but may not be sure. If the therapist brings it up, the client can feel hurt or reprimanded. If the therapist does not bring it up, the client can feel 'unsafe' with or have less respect for a therapist who she/he can easily deceive.

In the last 8 years of using e-therapy, I have had a couple of occasions where the client has accused me of 'not being fully present' with them on a Skype session. They have wondered if I am doing 'other things' on the internet, while also listening to them. Of course, if that were to be true, it would be highly unethical behaviour on my part. While, this accusation led to a fruitful discussion with one client, it led to an unfortunate termination with another.

With the first client, her accusing me of not listening to her fully led to a discussion of her carrying strong feelings of emotional neglect from the past. Her caregivers would often be doing other things in parallel, barely listening or tuning into her. When I told the client that I had not missed a single word she had said, and I was not doing anything else, she realized that she was projecting on to me some of the painful experiences of neglect from the past. Thereafter, we had a powerful session with her grieving the loss of emotional attunement from her primary caregivers.

Unfortunately, another client, with significant difficulties around trust and attachments had a tough time letting go of the doubt that I was doing something else while listening to him. It did not help that all his sessions with me were through videoconferencing and he could clearly see me on camera. My reassurances and explorations around the possibility of the reasons for his doubts did not go very far either and sessions eventually came to a stop. My sense is that this client had reached a point of closeness in the therapeutic relationship which was beginning to get uncomfortable for him and he probably needed to distance herself through this accusation, given that he had a history of being hurt in close relationships. As helpless and wrongly accused as I felt, I was forced to look at the clinical meaning of this experience. I could now understand why literature suggests that certain clients, especially the ones with fragile relationships/personality disorders, may not be the best candidates for online therapy. Face-to-face sessions might be more suited for them.

Clients and therapist encounters in the online space: Avoiding dual or multiple relationships with clients to be able to maintain neutrality in the therapeutic relationship is widely accepted in the practice of counselling and psychotherapy. It is also well understood that for most clients, it is not just helpful but essential to see their therapist only in the 'therapist role'. It can bring up a lot of difficult or conflicting emotions for a client to see their therapist outside of that role. While 'running into clients' at a mall, or a community/neighbourhood event, is a possibility all psychotherapists and counsellors have to be prepared for, mental health professionals who have any internet presence (for example, emails, social media sites, professional networking sites, online shopping) have to now also be mindful of the fact that information about them can be accessible to their clients through the internet. In this context, Zur (2008) points to the increasing incidence of accidental disclosures where the client accidentally bumps into information about the therapist online which she/he was not supposed to know. What complicates it even more is that when one 'runs into a client', at least one knows what has happened to be able to process it with the client in the next session. However, it would be hard for a therapist to know what the client has known about the therapist through information available online, unless the client shares it with the therapist.

The range of ethical situations varies from availability of information about the therapist in the online space to how the therapist negotiates limits of online non-clinical interaction with the client. The issues range from clients 'Googling' their psychotherapists (Kolmes 2010; Zur 2009), psychotherapists 'Googling' clients (Grohol 2010; Kolmes 2010; Zur 2010), to therapist online self-disclosure (Barnett 2010; Lehavot 2009; Younggren 2010; Zur et al. 2009) and therapists befriending a client on social media (Zur 2010).

These issues are relevant, and perhaps even inevitable, for all counsellors and therapists, regardless of whether they do e-therapy or not. While some are *digital immigrants* struggling to adjust to the steady rise of digitalism, others are *digital natives,* born into it and cannot imagine a life otherwise (Prensky and Berry 2001). Thus, the experience of navigating the online space with an ethical compass might differ between older therapists and the younger, comparatively *tech-savvy* trainees and early career therapists.

Therapists may have professional information about their qualifications, experience and contact information on their own website or on a LinkedIn profile. While this can be controlled, there may be more information about therapists available to clients if they search on the Internet. It is recommended that therapists periodically conduct an online search with their name to see what personal information or photographs might be visible publically.

Therapists might use status update services like Twitter or Blogs like Wordpress or Blogger to share their personal opinions and reactions to events. These may be accessible to clients and calls for caution regarding the content shared and an understanding of possible implications in the therapy space. A trainee therapist shared an unexpected ethical dilemma following her tweets about recent socio-political developments in the country. One of her clients commented on her tweet, expressing a completely different point of view, with a hostile tone. The trainee was uncertain about whether she should respond to the comment online, discuss it in the session or ignore it altogether, assuming that this was unrelated to the therapeutic process and relationship.

This situation illustrates the importance of reviewing the content of a personal tweet or post, assuming that it may be accessed by a potential or existing client. Adjustment of privacy settings and the use of pseudonyms could also be considered as options.

Experiences in a psychotherapy training context suggest that online encounters with clients throw up many questions. The exponential growth of social networking means that therapists, like others, may use this form of communication and interaction (Zur 2010). One of the most commonly experienced dilemmas that confront therapists relates to the appropriate response to a client's 'friend request' on a social networking site.

Any response to a 'friend request' should be a thoughtful process, which examines the impact on the boundaries of the relationship and the possible dual relationship across professional and personal spheres. The various questions that a therapist might ask oneself may be; What does this request mean to the client? What might be the implication of addressing the 'friend request'? What might be the possible impact on the client if the therapist declines the request? Should one simply 'ignore' the request or 'block' the sender? Is it okay to accept the friend request if therapist's social networking profile has very little personal information on it? Does the meaning still remain the same and do the boundaries still blur? How would the therapist have responded if the client asked during the session 'Will you be my friend?' The implications of accepting a 'friend request' on a social networking site are probably somewhat comparable to those that arise in a face-to-face session.

In response to this dilemma, most professional codes do not specifically mention what to do. However, organizations like American Psychological Association (2002), National Association of Social Workers (2008) and the American Counseling Association (2014) hold the view that multiple relationships should be avoided as much as possible, considering the potential of harm to the therapeutic relationship involved.

Some practitioners may respond to these anticipated difficulties with the stance of staying away from any social media networking sites. While this is one approach, others may choose to control the amount of information available to others by using privacy settings optimally. Privacy settings vary from having a 'limited profile' list, to making a content visible to selected few 'friends' and blocking visibility of 'public' content. The default settings on social media networking sites are dynamic and require frequent review. The American Medical Association (2010) also suggests separating personal and professional content online to maintain appropriate boundaries with the clients. This may be done by maintaining a separate social networking page sharing personal information and managing security settings accordingly, while also maintaining a separate social networking page for professional networking. This way the therapist will have control over the content shared and also who will be able to view the shared content on the social networking site. Preventive measures are recommended, rather than post hoc discussions after a 'friend request' has been sent. Discussions about the parameters of the therapist–client relationship, including online contact, are useful in the initial stages of therapy.

Just like the client can access information about the therapist through social media, the therapist might as well be tempted to gain more information about the client using the online forum. One trainee came in for a supervision session to discuss about a client with alcohol addiction issues who had missed therapy sessions for the past few months. Driven by curiosity, she did an online search using his name and found much evidence of a recent drinking binge. He came in for a session the following week claiming abstinence. The trainee was confused about whether she should tell him about the online search, confront him, remain silent, stop or continue to 'Google' the client in the future? Zur (2010) warns about the probable ethical concerns about impinging on a client's autonomy in such situations. An online search about a client might throw up information about the client which he may or may not intend to share with the therapist. What is more ethically troubling is, if in the process, the therapist comes across information that is therapeutically significant. The therapist might struggle with how to use the information and the merits and demerits of disclosure to the client. The recommendation is that online searches about a client should either be avoided. If this is done under exceptional circumstances or times of crisis, it should be a well thought action with specific intentions and considering any future implications (Kolmes and Taube 2014).

11.3.4 Personal and Institutional Social Media Policies

Mental health practitioners may develop a clear-cut personal social media policy to communicate to the client their expectations regarding social media interface with the client. This policy outlines the therapist's conduct on the Internet as a mental health professional and how a client can expect him/her to respond to various interactions that may occur between them on the Internet (Kolmes 2010). A typical social media policy may also highlight the changing nature of social media and a need to update the policy accordingly, assuring that an updated version will be provided to the client if required. The therapist may emphasise the fact that no form of interaction done online using any form of social media is completely secure, despite taking all available security measures. The specific details may include the stance related to 'friending' or allowing clients to 'follow' or be 'fans' on a social networking site, usage of social networking sites for interaction in between sessions or during emergency, including SMS and email and use of search engines.

Given the ubiquitous nature of digital media, many institutions are concerned about the detrimental impact of online actions by their trainees or employees. Wells et al. (2015) highlighted the need for an institutional social media policy with specific guidelines about a range of actions including posting client photographs online, the use of the institutional logo and online interactions with clients. Some may have specific indicators about social networking profiles and the acceptance of 'friend requests' from clients. For example, the Indiana University School of Medicine (2013) clearly indicates in its social media policy that a medical professional should not interact with the client online. The policy, particularly in a hospital set up, might be more specific in indicating if the therapist can keep in touch with the client for appointment-related purpose using emails or WhatsApp or similar forums. The specifics of the policy will depend on the type of institution and the philosophy it holds regarding therapy and the profession in general. The policies are developed primarily to protect clients, the institution and guide therapists at different stages in their professional journey.

11.4 Professional Guidelines and Digital Ethics

Standard ethical guidelines like that of American Psychological Association (2002) highlight the need for maintaining confidentiality and avoiding multiple relationships. However, the online presence of the client and therapist has redefined the substrate of the therapeutic interaction and requires interpretations of ethical codes from a different light, thus presenting more questions than answers.

This calls for the development of guidelines that specifically address the ethical concerns arising from digital interactions between therapist and client. The Indian Association of Clinical Psychologists (IACP 2015) has recently put forward a draft version of the ethical code which includes a brief section highlighting ethical concerns of e-therapy. This draft version advocates avoiding seeing clients with psychoses or eating disorder or if the presenting concerns are due to an immediate crisis, for e-therapy. It also mentions about the permanence of the digital trail, similar to other codes like the American Counseling Association (2014) and National Board for Certified Counselors (2016).

These codes urge practitioners to clearly indicate the limits of computer technology, and advise the therapist to use secured websites, email encryptions and other protective measures to secure online information as much as possible. The ethical guidelines of American Counseling Association (2014) and National Board for Certified Counselors (2016) discuss the need to include methods and duration of storage of session notes in the informed consent process and methods to identify client or therapist in an online forum.

The International Society for Mental Health Online (2009) and the American Medical Association (2010) are among the few guidelines specifically addressing online concerns. These also speak of providing the client with a schedule, response rate and alternate means of contacting the counsellor in the face of technological failure. The International Society for Mental Health Online emphasizes the need to provide emergency contacts in the locality. The American Medical Association (2010) advocates separate personal and professional pages on social networking sites. It is also suggested to avoid 'personal virtual relationships' with clients and to 'respect the privacy' of clients' social media unless given permission otherwise. They highlight that privacy settings should be used as much as possible on social networking sites to safeguard personal information and content. It should be realized that privacy settings are not absolute and that once on the Internet, content is likely to be there permanently. Thus, all the available codes unanimously recommend to monitor one's own Internet presence to ensure that the personal and professional information on their own sites and, to the extent possible, content posted about them by others, is accurate and appropriate. According to the American Medical Association (2010), if unprofessional content is posted by colleagues, a professional has the responsibility to bring that content to the attention of the individual, so that he or she can remove it and/or take other appropriate actions. If the behaviour significantly violates professional norms and the individual does not take appropriate action to resolve the situation, then the matter should be reported to appropriate authorities.

With the spread of e-therapy and its associated convenience, online therapy is being practised across states in a country, across countries and continents. The concern associated with this is whether a therapist licensed in a particular state/country can practise e-therapy with a client staying in a different state/country. With e-therapy as an emerging field, such issues are not yet adequately addressed in the professional codes or licensing bodies. In some states in the United States of America, licensed mental health professionals cannot practice online counselling in another state (Corey et al. 2014). In India, the situation is ambiguous with no definite guidelines indicating licensure limits of practising e-therapy across states or countries. One method of dealing with this uncertainty may be to check the licensure norms of the state or country the potential client belongs to, before initiating e-therapy.

11.5 Conclusion

The chapter has highlighted the ethical implications of online interface between client and therapist. Digitalism has engulfed all of us to varying degrees and has shown tremendous potential related to therapeutic service and professional development. While digitalism has brought both comfort and convenience, there are many fuzzy areas which are also surfacing. Experiences during workshops on digital ethics among health professionals, conducted by the second and third authors, highlighted uncertainties and prominent individual differences in the recognition and negotiation of ethical issues in the online space. There is a strong need for adequate training in this area and guidelines to help the therapist navigate through this apparent grey zone. However, the online space is an extremely promising platform for therapy and specific guidelines are also emerging. With little caution and technical knowledge, therapists can skilfully (and ethically) monitor their online presence and their interface with clients.

References

Alleman, J. R. (2002). Online counseling: The Internet and mental health treatment. *Psychotherapy: Theory, Research, Practice Training, 39*(2), 199–209.

American Counseling Association (2014). *American Counseling Association Code of Ethics*. Alexandria: Author.

American Medical Association (2010). American Medical Association Code of Ethics. Retrieved March 30, 2016 from http://www.ama-assn.org/ama/pub/physician-resources/medical-ethics/code-medical-ethics.page.

American Psychological Association (2013). Guidelines for the practice of telepsychology. Retrieved from http://www.apa.org/practice/guidelines/telepsychology.aspx.

American Psychological Association (2002). *Ethical principles of psychologists and code of conduct*. Washington: Author.

Barnett, J. E. (2010). Psychology's brave new world: psychotherapy in the digital age. *The Independent Practitioner, 30*(3), 149–152.
Bloom, J.W. (1997). NBCC Introduces its Standards for the Ethical Practice of Web Counseling. Retrieved from http://www.homemdemello.com.br/psicologia/nbcc.html.
Boyd, D. M., & Ellison, N. B. (2007). Social network sites: Definition, history, and scholarship. *Journal of Computer-Mediated Communication, 13*(1), 210–230. doi:10.1111/j.1083-6101.2007.00393.x.
Corey, G., Corey, M., Corey, C., & Callanan, P. (2014). *Issues and ethics in the helping professions with 2014 ACA codes*. Belmont: Cengage Learning.
Dewing, M. (2010). Social media: an introduction. *Library of Parliament, 1*, 1–5. Retrieved from http://www.lop.parl.gc.ca/content/lop/ResearchPublications/2010-03-e.pdf.
Frame, M. W. (1997). The ethics of counseling via the internet. *The Family Journal, 5*(4), 328–330. doi:10.1177/1066480797054009.
Grohol, J. (2010). Google and Facebook, therapists and client. [Web Log comment] Retrieved from http://psychcentral.com/blog/archives/2010/03/31/google-and-facebook-therapists-and-clients.
Indiana University School of Medicine (2013). *Guidelines for use of online social networks for medical students and physicians in training*. Retrieved from http://mse.medicine.iu.edu/files/shared/handbook/OnlineProfessionalism.pdf.
International Society of Mental Health Online (2009). *Suggested principles for the online provision of mental health services*. Retrieved from http://www.ismho.org/suggestions.asp.
Kanani, K., & Regehr, C. (2003). Clinical, ethical, and legal issues in e-therapy. *Families in Society: The Journal of Contemporary Human Services, 84*(2), 155–162. doi:10.1606/1044-3894.98.
Kolmes, K. (2010). Developing my private practice social media policy. *The Independent Practitioner, 30*(3), 140–142.
Kolmes, K., & Taube, D. O. (2014). Seeking and finding our clients on the Internet: Boundary considerations in cyberspace. *Professional Psychology: Research and Practice, 45*(1), 3–10.
Lee, S. (2010). Contemporary issues of ethical e-therapy. *Journal of Ethics in Mental Health, 5*(1), 1–5.
Lehavot, K. (2009). "MySpace" or yours? The ethical dilemma of graduate students' personal lives on the internet. *Ethics and Behavior, 19*(2), 129–141.
Maheu, M. M., Pulier, M. L., Wilhelm, F. H., McMenamin, J. P., & Brown-Connolly, N. E. (2005). *The mental health professional and the new technologies: A handbook for practice today*. New Jersey: Taylor & Francis.
Mallen, M. J., Vogel, D. L., & Rochlen, A. B. (2005). The practical aspects of online counseling-ethics, training, technology, and competency. *The Counseling Psychologist, 33*(6), 776–818.
Manhal-Baugas, M. (2001). E-therapy: practical, ethical, and legal issues. *CyberPsychology & Behavior, 4*(5), 551–563.
McWilliams, N. (2004). *Psychoanalytic psychotherapy: a practitioner's guide*. New York: The Guilford Press.
National Association of Social Workers (2008). *Code of ethics of the national association of social workers*. Retrieved from http://www.naswdc.org/pubs/code/code.asp.
National Board for Certified Counselors (2016). Policy regarding the provision of distance professional services. Retrieved from http://www.nbcc.org/Assets/Ethics/NBCCPolicyRegardingPracticeofDistanceCounselingBoard.pdf.
Prensky, M., & Berry, B. D. (2001). Do they really *think* differently? *On the Horizon, 9*(6), 1–9.
Ragusea, A. S., & VandeCreek, L. (2003). Suggestions for the ethical practice of online psychotherapy. *Psychotherapy: Theory, Research, Practice Training, 40*(1–2), 94–102. doi:10.1037/0033-3204.40.1-2.94.
Suler, J. (2001). Assessing a person's suitability for online therapy: the ISMHO clinical case study group. *CyberPsychology & Behavior, 4*(6), 675–679.

Wells, D. M., Lehavot, K., & Isaac, M. L. (2015). Sounding off on social media: the ethics of patient storytelling in the modern era. *Academic Medicine, 90*(8), 1015–1019.
Winnicott, D.W. (1960). Ego distortion in terms of true and false self. *The maturational process and the facilitating environment* (pp. 140–152). New York: International Universities Press.
Younggren, J. N. (2010). To tweet or not to tweet, that is the question. *The Clinical Psychologist, 63*(2), 18–19.
Zack, J. S. (2004). The technology of online counseling. In R. Kraus, J. Zack, & G. Stricker (Eds.), *Online counseling: a handbook for mental health professionals* (pp. 93–121). San Diego: Elsevier Academic Press.
Zur, O. (2008). The Google factor: psychotherapists' self-disclosure in the age of the internet: discover what your clients can find out about you with a click of the mouse. *The Independent Practitioner, 28*(2), 82–85.
Zur, O., Williams, M. H., Lehavot, K., & Knapp, S. (2009). Psychotherapist self-disclosure and transparency in the internet age. *Professional Psychology: Research and Practice, 40*(1), 22–30.
Zur, O. (2010). To Google or not to Google. our clients? When psychotherapists and other mental health care providers search their clients on the web. *Independent Practitioner, 30*(3), 144–148.

Chapter 12
Ethics in Therapeutic Practice: Culturally Universal and Valid?

R. Raguram

Abstract The cardinal principles of ethics in the practice of psychotherapy comprising of autonomy, beneficence, justice and non-maleficence are well known. This paper advances the position that ethical principles must be viewed within the prism of sociocultural forces that shape and influence them rather than as universal concepts that are invariant. It attempts to address the challenges that arise when therapists try to operationalize these principles in their practice in the Indian context. It is argued that ethical principles are emergent properties of interactions between the therapists and their clients, with no a priori assumptions and are closely linked to the social structures of which both are a part. Impoverishment and exploitation (both economically and emotionally) are intrinsic to the sociocultural milieu just as much as internal dynamic conflicts are. In this scenario, ethics cannot be decontextualized from local realities. Psychotherapeutic interventions have to be sensitive to these processes to be culturally congruent and socially relevant.

Keywords Ethics · Decontexualization · Cultural sensitivity · Social relevance · Psychotherapy

The field of biomedical ethics has been enormously influenced, both in theory and in practice, by the four principles approach, which has its origins in the 1978 *Belmont Report* and the work of Thomas Beauchamp and James Childress (2006). According to these models, a physician's moral responsibility towards his or her patient is defined by four ethical principles—respect for autonomy, non-maleficence, beneficence and justice. Respect for autonomy dictates that those patients who have decision-making capacity have a right to voice their treatment preferences, and physicians have the attendant duty to respect those preferences. Non-maleficence directs physicians to maximize the benefit to patients while minimizing the harm. Beneficence promotes the welfare and best interest of

R. Raguram (✉)
Department of Psychiatry, Kempegowda Institute of Medical Sciences and Research, Bengaluru, India
e-mail: profraguram@gmail.com

P. Bhola and A. Raguram (eds.), *Ethical Issues in Counselling and Psychotherapy Practice*, DOI 10.1007/978-981-10-1808-4_12

patients. Finally, justice demands fair, equitable and appropriate treatment for all patients. These ethical principles are commonly referred to in most professional ethical guidelines and applied in clinical decision-making. It is argued that this approach offers a common, basic moral analytical framework and a common, basic moral language (Gillon 1994). The moot question, however, is whether these principles are culturally neutral and can be applied inflexibly in diverse sociocultural settings. The relevance and applicability of this framework needs closer scrutiny especially with reference to psychotherapy as it is practised in the Indian setting.

12.1 Autonomy

The concept of autonomy has its origins in the Western culture, which emphasizes individualism, personal happiness and self-actualization. In this context, 'personhood' is viewed from the perspective of autonomy and individual rights. In many Western cultures, there is an abiding emphasis in the inherent separateness of individuals. The normative quest in these settings is for the person to individuate and to explore one's unique aspirations and objectives in life.

As a consequence, the overarching emphasis is on actualizing one's own aspirations in preference to the thoughts, feelings and actions of others. As Geertz (1975) remarked, the person is viewed as "a bounded, unique, more or less integrated motivational and cognitive universe, a dynamic centre of awareness, emotion, judgement and action organized into a distinctive whole and set contrastively both against other such wholes and against a social and natural background". However, researchers have called into question the Western view of the individual as an independent, self-contained autonomous entity (Marsella et al. 1985). For instance, it has been suggested that in some cultures, the individual, in the sense of a set of significant inner attributes of the person, may cease to be the primary unit of consciousness. It entails seeing oneself as part of an encompassing social universe and recognizing that one's behaviour is influenced and organized to a significant extent by the thoughts, feelings and actions of others (Sampson 1988).

It is imperative to examine how the concept of autonomy is actually translated in real life situations, especially in the Indian setting. Based on his extensive, ethnographic field work in Orissa, Schweder et al. (1997) drew attention to the multifaceted notion of ethics often referred to as the "Big Three" ethics of autonomy, community and divinity. The ethics of autonomy is based on people's right to pursue their needs and desires. On the other hand, the ethics of community pays attention to actions consistent with one's social role. The ethics of divinity is anchored in the concept of divine or natural law, often based on religious authorities and texts which connect the individual aspirations to a higher force. It is pertinent to note that these three facets interweave with each other imperceptibly. Further, each culture emphasizes a particular set of principles, situating them in the wider context of that culture. For example, it is suggested that Americans draw from the ethics of autonomy more than Brazilian, Indians or Filipinos and conservative Christians and Orthodox Jews rely on the ethic of divinity (Jensen 2008).

Through an exceptionally rich and detailed ethnographic account of the Bakkarwal, a Muslim pastoralist community of Kashmir, Rao (1998) set out to explore the question of the extent to which the exercise of individual autonomy or agency is available to Bakkarwal women and men as they progress through successive life stages. Her work constitutes an empirical investigation of local Bakkarwal concepts of "autonomy", as it is manifested in day-to-day talk and social interaction. Through extensive use of indigenous concepts, she delineates as to how the concept of autonomy is influenced to a large extent by gender. While autonomy and decision-making was the prerogative of the men, it was elusive for women as they had to endure "extreme lack of autonomy".

This study draws attention to an often ignored dimension of gender roles in the conception of autonomy. Autonomy is not entirely predicated within the individual, but is relational. The concept of relational autonomy underscores the fact that autonomy is not achieved by simply choosing from among an array of options. It requires us to examine what are the options on offer, how these have arisen and crucially, why certain options are not available or accessible in situations where choices are constrained by unequal and hierarchical power relations (Sherwin 2008). Gendered power relations pose several challenges in psychotherapy since the self is not merely an exclusive privatized entity operating within a closed system. When the therapist approaches the client from a relational autonomy perspective, the focus is on a relational being with a socio-historical context. The task is to attend to the needs of the individual self which is encompassed within one or many other selves. The client is always a person-in-relation. The therapeutic orientation must strive to achieve a balance between the traditional social positions of self on the one hand, and respect for individual rights and autonomy on the other, while holding them together in productive tension.

In terms of ethics, patient autonomy in Indian culture may not play the same role that it does in the West. In practical terms, autonomy may mean different things in different cultural and geographic settings, or at least it can have different applications. The issue is not about patient autonomy but about broader relationships between personal autonomy, family and society in the Indian setting.

12.2 Beneficence and Non-Maleficence

The term *beneficence* is broadly understood in ethical theory to include virtually all forms of action intended to benefit or promote the good of other persons. It implies a moral obligation to act for the others' benefit, helping them to further their important and legitimate interests, often by preventing or removing possible harms. The social sanction for psychotherapy is based on the notion that it does good for the client. Tjeltveit (2006) argued that the notion of beneficence rests on value judgements about what constitutes a good outcome, as opposed to a bad one. Let me amplify this aspect by a case vignette (Case 12.1).

Case 12.1
Mr. V was referred by a medical practitioner with a note stating that he was depressed and had failed to improve with antidepressants and might do well with cognitive behaviour therapy. Mr V was a 43-year-old agriculturist from a rural region of Karnataka. He had a small piece of land in which he was cultivating cotton. He had taken considerable sums of money from a local money lender to buy seeds and fertilizers. In that particular year, many farmers like him were compelled to buy genetically modified seeds which were also costlier than the regular ones. They needed more irrigation. Unfortunately the monsoon played truant and the seeds failed to yield the expected output. The entire crop was a failure. The money lender started exerting pressure for speedy return of the loan and when he was tardy in repaying it, took possession of the original documents of his property. It was a tough period in his life as he had no source of income other than his agricultural produce. There was hardly anything to sustain the family. He felt terribly sad about the dire situation his family was facing and had lost all hopes of overcoming the crisis. An overwhelming sense of helplessness loomed over him. He had lost all interest in life and the future; however acting on a friend's suggestion he met the doctor who had started him on an antidepressant.

As he was sharing his distress with much agony, it struck me that what could be construed as "negative cognitions" in terms of helplessness and hopelessness were in fact accurate representations of his disturbing everyday life situation. I wondered whether by viewing them as "cognitive distortions" and initiating the process of 'correcting' them, was I covertly facilitating an adaptation to an unliveable situation instead of enabling him to confront the oppressive forces of which he was a victim? More importantly, what would be construed as a 'good' outcome in such situations where individual life is constrained by existing circumstances which limit opportunities for an authentic existence? Beneficence should also aim at protecting and empowering the rights and opportunities of clients, rather than limiting itself to addressing issues within the individual which might in fact disempower them.

12.3 Justice

Justice is one of the fundamental concepts impacting the quality of human relationships. The principle of justice in psychotherapy requires that the therapist is just and fair to all clients, respecting their dignity and human rights. Often the inner conflicts of clients occur in the context of unsettling social deprivation which constrain the individual. It is imperative in such circumstances to work towards establishing new frameworks of justice. Let me amplify this issue through a case illustration (Case 12.2).

Case 12.2

A 25-year-old patient was brought to my consultation room. He was dressed shabbily in torn clothes and his hair was matted and dirty. There were bruises on his body, especially on his limbs, suggesting that he had been tied up with ropes for a while. He had a vacant look on his face and remained curled up in the hospital bed. He did not respond to verbal queries and when an attempt was made to touch him, he went into violent convulsions, as if hitting out at imaginary people. The people in whose house he was working reported that he had suddenly began to have similar 'convulsions' a few days before and that was the reason for bringing him to the hospital. They said that they had to tie him so that he would not hurt himself during the convulsions. He had also been refusing food and had stopped speaking.

Over the next few days, I gathered some details about his background and his current work situation from a co-worker who stayed with him in the hospital. The client was the oldest son and had four unmarried sisters, all of whom lived in his native village. His father was 65 years old and was an alcoholic. Though he had an arable land, he was inconsistent in work because of his alcohol use and had borrowed heavily from a local landlord. Over several years, the loan amount kept mounting and he was not in a position to pay even the interest, which had accumulated to a considerable extent. As the principal and the interest mounted, the landlord took possession of the land and also took away the client to work in his farm. The client was not paid any wages as it was adjusted towards the loan amount. The work was quite heavy and the demands on the client were many. He was also not allowed to meet his family. A year later, when the client's mother died, the landlord took him to the funeral but brought him back the same day. Over the years, the client's contact with his family had declined greatly and he had shared his distress concerning this with his co-workers. A few days prior to the onset of the current problem, the client requested the landlord to permit him to visit his family as his father was reportedly unwell. The landlord got extremely angry and beat him badly. The next day, the client did not go to work, saying that he felt very weak and tired. This further infuriated the landlord and more physical abuse followed. Later, on the same day, the client started having 'convulsions' and stopped talking and taking food. The local doctor who examined him advised that the patient be taken to the hospital. The examining clinician in the emergency ward opined that these were 'pseudoseizures' and hence the client was referred to the psychiatry department.

After admission, the client was kept on intravenous fluids as he refused to eat. Over the next few days, his general condition improved and he started talking. His first utterance was "please do not hurt me further" and he started crying. When he was reassured, his response was "Will you make my pain go away....for ever?" He became more communicative over the next few days and did not have any 'convulsions'. He gradually started talking about the problems he was facing at work: "I am a bonded labourer. I have been working with this landlord for several years as my father was unable to repay the loan. He doesn't give me any wages and doesn't allow me to meet my family. He beats me often. I wonder if he will ever free me. If he comes to know that I have said all this to you, he will beat me more when I go back. Please do not send me back to him".

What would be an appropriate framework to respond to the client's plight? Clinically what he was suffering from can be termed as 'pseudoseizures' or "conversion disorder". Classical psychodynamic understanding would attribute his 'convulsions' to the repressed conflicts concerning his employer which are then modulated by the defence mechanism of 'conversion' as violent motor phenomena. Facilitating ventilation and providing an opportunity to share his disturbed feelings would no doubt reduce his psychological distress. But what would be a 'just' response in such situations? Should the therapist focus attention exclusively on internal psychodynamics in isolation of the external power dynamics? Psychotherapy rarely extends its ambit to address the latter. There is great scepticism about the potential of psychotherapy as an instrument to facilitate social justice even among practitioners, many of whom see what they do as contributing to individual healing from personal trauma but not as an instrument of liberation from societal oppressions. Often psychotherapists' emphasis on "adjustment" impedes the process of empowerment necessary for true healing. When internalized societal inequalities and traumatic experiences impact the client's mental health through oppression, the therapist should help the client to recognize how the two are linked. The impact of internalized societal oppressions on the client's unconscious cannot be ignored and the emphasis in therapy should shift from self-defeating privatization of difficulties.

Psychologist Ignacio Martín-Baró, who was one of the six Jesuit priests murdered by government-sponsored assassins in El Salvador on 16 November 1989, contended that

> ... "It is not in the hands of psychologists, to change the structural socioeconomic injustice in our countries. Nevertheless, there is important work for psychologists to do. If it is not the calling of the psychologist to intervene in the socioeconomic mechanisms that cement the structures of injustice it is within the psychologist's purview to intervene in the subjective processes that sustain those structures of injustice and make them viable (Martín-Baró Ignacio 1994).

Franz Fanon, practising in the mental health centres of French-dominated Algeria, stated more forcefully his firm conviction;

> "that therapy should, above all, restore freedom and justice to patients.... Psychological work with the oppressed which is not about dis-alienation of praxis and regaining of power

tends to produce morally entrapped and compromised *objects*, not liberated and creative *subjects*" (cited in Bulhan 1985).

Yet, currently, most practitioners of psychotherapy do not emphasize social justice *within* their practice *as psychotherapists*. Many of us are still unsure of how to intervene "in the subjective process" of individuals through psychotherapy so that it can address the "social structures of injustice" without compromising our ethics and professional skills. To begin with, we must become convinced that therapy ought to include awareness of issues of social justice in the client's context. This means recognizing that the degree of "psychological integration of the individual… is profoundly affected by the social milieu in which the individual lives…" (Pizer and Travers 1975).

Therapy is often deemed to be an insular process aimed at effecting change within the individual. At the same time, the therapist is also a social being. This places therapists in a crucially important yet contradictory relationship with justice. However the principle of justice as it is operative in psychotherapy is seldom discussed or debated.

Therapists need to evolve and execute two functions: looking inwards to work through emotional issues within the individual while also empowering the client to work through unequal power dynamics in life and facilitating a socially relevant ethics of justice.

McWhirter (1991) puts it succinctly: "clients should be viewed *not* as 'sick' people in the system, but as people attempting to cope as best they can within a 'sick' system". Psychotherapy's reluctance to address social power dynamics which are as salient as internal psychodynamics, seriously impacts a just therapeutic practice.

12.4 Non-Maleficence

The principle of non-maleficence emphasizes an obligation on the part of the therapist not to inflict harm intentionally. Therapists must refrain from providing ineffective treatments with little benefits or those that might harm the interests of the clients. It also involves avoiding any kind of exploitation of the client, emotionally or sexually.

There are three different and overlapping components to harm; negligence and malpractice, misconduct and imposition of personal values and prejudice. Negligence and malpractice contribute to errors in practice. Like any other therapeutic initiative, psychotherapeutic process is not immune to errors. But strangely, this aspect has been rarely addressed in psychotherapeutic literature. Medau et al. (2013) provide a useful framework to examine errors in psychotherapy. They classify psychotherapeutic errors into technical, judgemental, normative and system errors. Technical errors concern the procedural aspects of therapy. Judgemental errors refer to the improper choice and use of therapeutic methods. When issues of confidentiality are breached, it is considered a normative error. System errors refer to improper or inadequate practices like lack of time spent in therapy or inadequate follow-ups. This offers a useful framework to examine and address errors to ensure that adverse consequences in therapy are minimized. Acknowledging that errors do occur in psychotherapy and addressing them effectively will contribute significantly to mitigate issues of maleficence.

Imposition of personal values and prejudice can be minimized through appropriate training and mandatory supervision as it involves issues of therapists' competence. The American Psychological Association's Ethical Principles of Psychologists and Code of Conduct (1992) clearly enunciates the issues of competence. It states that "psychologists provide services, teach, and conduct research with populations and in areas only within the boundaries of their competence, based on their education, training, supervised experience, consultation, study, or professional experience". It is imperative that psychotherapists undergo appropriate training supplemented by rigorous supervision to ensure quality of practice which does not adversely impact the client. Unfortunately this is a distant dream in the Indian context. Very few centres offer supervised training in psychotherapy. There is hardly any emphasis or requirement for continued supervision after the training period.

Misconduct in therapy can significantly harm the interests of the client. Identifying misconduct and addressing them would involve institutional safeguards. Regretfully there is no mandatory process of certification and registration of psychotherapists in India. This has resulted in a profusion of inadequately trained and improperly supervised therapists, a process that contributes significantly to the issue of non-maleficence.

12.5 Conclusion

The conventional four principles approach of Beauchamp and Childress (2006) is steeped in common morality. But moral norms spring from particular cultural, social and institutional sources. It must also be borne in mind that these four principles lack any systematic relationship to each other and there is no unified moral theory from which they are all derived. Moreover, moral issues are embedded in a contextual frame that eludes abstract, universal reasoning. Reasoning about ethical principles based on overarching moral rules diverts attention from the complexities of everyday life as lived and experienced by clients. What matters in a particular therapeutic encounter are the situational specifics in which we find ourselves—our own history, the particulars of the client with whom we are dealing and the implicit power dynamics.

Ethics opens up the interplay between culture and individual consciousness. Psychotherapy as it is practised in the Indian context is essentially a post-colonial discourse influenced by Western paradigms and concepts. It focuses attention on the circumscribed 'psyche' of the individual by adopting a detached expert position. Culturally congruent therapeutic processes necessitate an understanding of the client as a socially and relationally embedded being, subject to discrimination and injustice. By adopting this stance, the therapist will be able to accept and validate the client as a unique being subject to a flux of distinctive experiences. This humanizing ethic is eloquently articulated in Buber's (1958) concept of the 'I-Thou' attitude. Buber contrasts this with the 'I-It' attitude, in which the person is experienced as an objective entity that can be systematized, analysed and broken down into universal parts. It is through this perspective to respect and validate the client in the totality of their being that we acknowledge, validate and work through the myriad social, economic and cultural limitations that they might face in their lives.

Ethical issues must be viewed within the prism of sociocultural forces that shape and influence them rather than as a concept that is invariant. When we begin to consider ethical codes as Ultimate Truths, we tend to not pay attention to how they have to be refined and adapted to contextual needs and determinants. In the process, we fail to recognize that ethical principles are socially constructed. An individual's ethical landscape is inextricably linked to the relationships in which s/he exists and the preservation of that connectedness is a crucial component in ethical decision-making (Cottone 2001). This framework envisages that ethical decisions are always interactional, operating within a consensual domain. Recognizing this is imperative for a culturally sensitive practice of psychotherapy.

The essence of ethics springs from the pursuit of the good—in conduct, in personal relationships, in how we treat those around us. It is unethical to blindly follow ethical codes which are deemed universal without scrutiny of their content, their values, and their assumptions in particular sociocultural contexts. Ethical codes are, after all, socially shared patterns of moral judgements and behaviour. Hence it is important to find a more flexible and broader concept of ethics that lends itself for adaptation in varying contexts. Lindsay et al. (2008) enunciated a set of ethical principles to specifically assist the therapists in their work in Europe. This encompasses respect, competence, responsibility and integrity. Respect is divided into general respect, privacy, informed consent and confidentiality. The principle of competence describes ethical awareness and knowledge. Knowing the limits of one's own competence and taking advanced training are also subsumed under this principle.

The principle of integrity describes recognition of professional limitations through self-reflection, honesty and accuracy. The principle of responsibility calls for the best possible treatment and the avoidance of harm. The Lindsay approach has been adopted by the Meta-Code of Ethics of the European Federation of Professional Psychological Associations (2005) as well as by the British Psychological Society (2009). This framework is an attractive alternative to explore in the Indian context as it offers a wider perspective than the conventional four principle approach to ethics.

Ethical guidelines have been largely aspirational in their orientation. The critical problem is to translate them to socioculturally resonant praxis since in the final analysis ethics must be practical (Pope and Vasquez 1998). Therapists practising in the Indian context confront an almost unimaginable diversity of situations, each with its own shifting questions, demands, and responsibilities. Ethics that are out of touch with the practical realities of therapeutic discourse would be of limited value. Hence it is important that we cast our gaze beyond our consulting rooms to the diversity of worldviews outside, which impact and influence our clients in their everyday lives, to ensure that our therapeutic initiatives are practically relevant and personally meaningful.

References

American Psychological Association. (1992). Ethical principles of psychologists and code of conduct. *American Psychologist, 47*, 1597–1611.

Beauchamp, T. L., & Childress, J. F. (2006). *Principles of biomedical ethics* (5th ed.). New York: Oxford University Press.

Bowman, K. (2004). What are the limits of bioethics in a culturally pluralistic society? *Journal of Law and Medical Ethics, 32*(4), 664–669.

Buber, M. (1958). *I and Thou (R.G. Smith, Trans* (2nd ed.). Edinburgh: T & T Clark Ltd.

Bulhan, H. A. (1985). *Franz Fanon and the Psychology of Oppression*. New York: Plenum Press.

Cottone, R. R. (2001). A social constructivism model of ethical decision making in counselling. *Journal of Counselling and Development, 79*(1), 39–45.

European Federation of Psychologists' Association. (2005). *Meta code of ethics: guidance for ethical codes of member associations*. Retrieved on 16th April 2016 from http://wwe.efpa.eu/ethics/ethical-codes.com.

Geertz, C. (1974). From the native's point of view: On the nature of anthropological understanding. In K. Basso & H. Selby (Eds.), *Meaning in anthropology* (pp. 221–237). Albuquerque: University of New Mexico Press.

Geertz, C. (1975). On the nature of anthropological understanding. *American Scientist, 63*(1), 47–53.

Gillon, R. (1994). Medical ethics: four principles plus attention to scope. *British Medical Journal, 309*(6948), 184–188.

Jensen, L. A. (2008). Through Two Lenses: A Cultural-Developmental Approach to Moral Psychology. *Developmental Review, 28*(3), 289–315.

Lindsay, G., Koene, C., Ovreeide, H., & Lang, F. (2008). *Ethics for European psychologists*. Göttingen: Hogrefe.

Marsella, A., De Vos, G., & Hsu, F. L. K. (1985). *Culture and self*. London: Tavistock.

Martín-Baró, I. (1994). The role of the psychologist. In Ignacio Martín-Baró, A. Aron & S. Corne (Eds.). *Writings for a Liberation Psychology* (pp.33–46). Cambridge, MA: Harvard University Press.

McWhirter, E. H. (1991). Empowerment in counselling. *Journal of Counselling and Development, 69*, 222–227.

Medau, I., Jox, R. J., & Reiter-Theil, S. (2013). How psychotherapists handle treatment errors: an ethical analysis. *BMC Medical Ethics, 14*, 50.

Pizer, Stuart A., & Travers, Jeffrey R. (1975). *Psychology and Social Change*. New York: McGraw-Hill.

Pope, K., & Vasquez, M. (1998). *Ethics in psychotherapy & counseling: A practical guide* (2nd ed.). San Francisco: Jossey-Bass.

Rao, A. (1998). *Autonomy. Life Cycle, Gender and Status among Himalayan Pastoralists*. New York-Oxford: Berghahn Books.

Sampson, E. E. (1988). The debate on individualism: Indigenous psychologies of the individual and their role in personal and societal functioning. *American Psychologist, 43*(1), 15–22.

Sherwin, S. (2008). Whither bioethics? How feminism can help reorient bioethics. *International journal of feminist approaches to bioethics, 1*(1), 7–27.

Shweder, R. A. (1990). In Defense of Moral Realism: Reply to Gabennesch. *Child Development, 61*(6), 2060–2067.

Shweder, R. A., Much, N. C., Mahapatra, M., & Park, L. (1997). The "Big Three" of Morality Autonomy, Community, Divinity and the "Big Three" Explanations of Suffering. In A. Brandt & P. Rozin (Eds.), *Morality and Health* (pp. 119–169). New York: Routledge.

Tjeltveit, A. C. (2006). To what ends? Psychotherapy goals and out-comes, the good life, and the principle of beneficence. *Psychotherapy: Theory, Research, Practice. Training, 43*(2), 186–200.

The National Commission for Protection of Human Subjects of Research (1979). *The Belmont Report—Ethical Principles and Guidelines for Protection of Human Subjects of Research.* Retrieved on 30th March 2016 from http://www.hhs.gov/ohrp/humansubjects/guidance/belmont.html.

The British Psychological Society. (2009). Code of ethics and conduct. Retrieved on 16th April 2016 from http:// www.bps.org.uk/system/files/documents/code_of_ethics_and_conduct.pdf.

Chapter 13
Ethical and Legal Issues in Psychotherapy

B. Vinay, J. Lakshmi and Suresh Bada Math

Abstract The chapter presents an overview of ethical and legal issues pertaining to the clinical practice of psychotherapy. Psychotherapy is an English word of Greek origin, deriving from the ancient Greek words 'psyche' (ψυχή meaning 'breath; spirit; soul') and 'therapia' (θεραπεία meaning 'healing; medical treatment'). Psychotherapy is a science which alleviates pain, suffering, stress or anxiety. Psychotherapy may also be used in combination with medicines to treat mental illness. Ethical and legal issues which are likely to be encountered in practice; like record keeping, confidentiality, boundary crossings and violations, medical negligence and other practical issues have been examined in this chapter. The ability to think rationally and apply general ethical principles to specific situations is vital. The purpose of this chapter is to provide an understanding of ethical and legal issues related to psychotherapy. However, it does not replace advice based on legal expertise.

Keywords Ethics · Legal · Psychotherapy

Psychotherapy is a general term for treating mental health problems by talking with a psychiatrist, psychologist or other mental health service providers. However, a precise definition of psychotherapy is, "Treatment, by psychological means, of problems

---

B. Vinay
Department of Psychiatry, The Oxford Medical College & Research Center, Bengaluru 562107, India
e-mail: bvinay93@gmail.com

J. Lakshmi
Department of Clinical Psychology, National Institute of Mental Health and Neuro Sciences, Bengaluru 560029, India
e-mail: lakshminair999@gmail.com

S.B. Math (✉)
Department of Psychiatry, National Institute of Mental Health and Neuro Sciences, Bengaluru 560029, India
e-mail: sbm@nimhans.kar.nic.in; nimhans@gmail.com

P. Bhola and A. Raguram (eds.), *Ethical Issues in Counselling and Psychotherapy Practice*, DOI 10.1007/978-981-10-1808-4_13

of an emotional nature in which a trained person deliberately establishes a professional relationship with the patient with the objective of removing, modifying or retarding existing symptoms; mediating disturbed patterns of behaviour and promoting positive personality growth and development" (Wolberg 1988).

There are different types of psychotherapies such as, psychoanalytic therapy, cognitive behavioural therapy, family therapy, interpersonal therapy, supportive therapy, eclectic therapy and brief therapies. These therapies have have varied underlying theories, approaches, target ingredients, durations, modalities and may be recommended for specific populations. However, these therapies show that there are certain ingredients which are common to all kinds of psychotherapy. Psychotherapy includes interactive processes between a person or a group; and a qualified mental health professional. The goal of psychotherapy is to alleviate pain, suffering, stress or emotional distress through exploration of thoughts, feelings and behaviour (Wolberg 1951). Psychotherapy is an agreement between a practitioner (usually called 'Psychotherapist') and a client to work together towards alleviating pain by entering into a therapeutic relationship (Sills 2006). By this agreement, the therapist and the client form a therapist–client relationship. Once this relationship is established, the therapist has a duty of care to the client. Guidelines have been established to maintain minimum standards of ethical practice. Considering the dynamic interaction between the practitioner and the client, there are many ethical and legal issues that arise during psychotherapy (Norcross et al. 2011; Wolberg 1988).

When a professional psychotherapist agrees to attend to a patient, there is an unwritten contract between the two (Croxton 1985; Sulzer 1962). The contract can be defined as an agreement between two or more persons which creates an obligation between them, to do or not to do a particular thing. The contract may be expressed or implied. The patient entrusts himself to the therapist and the therapist agrees to do his/her best, at all times, for the patient (Croxton 1985). The therapist–patient contract is almost always an implied contract, except where a written informed consent is obtained. Under the therapist–patient contract relationship there are certain obligations for the therapist such as to continue treatment, provide reasonable care within reasonable skill, and not to undertake any procedure beyond his/her skill and professional secrets (Feldman and Ward 1979; Simon 1991; Van Hoose and Kottler 1985). Increasing legal and regulatory demands on 'health-care practitioners', including demands that directly conflict with the central tenets of traditional professional ethics (such as confidentiality), were a major factor in motivating the formation of the professional associations to form minimum standards for care, licensing, remuneration and ethical guidelines (Simon 1991; Van Hoose and Kottler 1985). The industrialization of health care and information technology revolution has added more complexities in this therapeutic relationship. Consequently the proliferation of guidelines, ethical issues, legislation designed to 'protect the client/public' were greatly limiting psychotherapists' and psychoanalysts' clinical freedom (Austin et al. 1990; Schutz 1982; Thompson 1983). To overcome these barriers, this chapter discusses ethical and legal issues involved in psychotherapy practice and also attempts to provide general guidelines for safe practice. The case vignettes used in the chapter are fictitious and are used to illustrate common dilemmas. Further, this chapter does not intend to serve in lieu of clinical, ethical and legal consultation. It neither gives legal, ethical nor clinical advice. In this dynamic world, each situation, context, therapist and client is different, and, therefore, each situation needs to be evaluated independently.

Case 13.1
Mr. X, 33 years, an unmarried advocate, from middle socio-economic status, diagnosed with depression and personality disorder, was on treatment from a well-known hospital. He was on medication and also on psychotherapy. He had already completed 24 sessions of psychotherapy. During the psychotherapy sessions, he developed transference towards the therapist and started making frequent phone calls, sent messages on the mobile phone and wrote emails. The therapist worked on handling the transference and tried to explain the boundaries in the professional relationship but he became further depressed and stopped going to work. Finally the therapist terminated the sessions. This resulted in immense distress and pushed Mr X to commit suicide.

Did the psychotherapist fulfil the therapeutic obligation of providing care? Was the psychotherapist negligent in terminating the sessions prematurely? What are the ethical and legal ramifications in this case?

13.1 Professional Negligence

Psychotherapy is a well-known profession in which the practitioner provides mental health care to the suffering (Jenkins 2002). Professional negligence, occurs when a person, in the course of practicing his or her profession, improperly performs the duties of that profession, and someone is harmed as a result (Bernstein and Hartsell 2004; Jenkins 2002; Packman et al. 2004) see Case 13.1.

The Black Law dictionary definition of negligence is:

> … conduct, whether of action or omission, which may be declared and treated as negligence without any argument or proof as to the particular surrounding circumstances, either because it is in violation of statue or valid municipal ordinance or because it is so palpably opposed to the dictates of common prudence that it can be said without hesitation or doubt that no careful person would have been guilty of it. As a general rule, the violation of a public duty, enjoined by law for the protection of person or property, so constitutes.

In simple terminology ‘Negligence’ means ‘Lack of Proper Care’.

The Supreme Court of India discussed the conduct of professionals and what may amount to negligence by professionals in Jacob Mathew v. State of Punjab (2005):

> In the law of negligence, professionals such as lawyers, doctors, therapist, architects and others are included in the category of persons professing some special skill or skilled persons generally. Any task which is required to be performed with a special skill would generally be admitted or undertaken to be performed only if the person possesses the requisite skill for performing that task. Any reasonable man entering into a profession which requires a particular level of learning to be called a professional of that branch, impliedly assures the person dealing with him that the skill which he professes to possess shall be exercised and exercised with reasonable degree of care and caution.

Various professionals have complained about the large number of litigation suits and have urged legal reforms to curb large damage awards, whereas tort attorneys have argued that negligence suits are an effective way of compensating victims of negligence and of policing the profession. Professional negligence can be best understood by studying ‘Medical Negligence’, because there are large numbers of case laws available across the globe. When there is a breach in a contractual obligation of providing care, with a breach of duty which invites the intervention of judges to grant certain remedies for the damages, then the tortious liability arises.

A person who alleges negligent medical malpractice must prove the following elements:

1. A duty of duty care: a duty exists whenever a hospital or healthcare provider or therapist undertakes care or treatment of a patient.
2. Breach of duty: the provider failed to conform to the relevant standard care.
3. The breach caused an injury/damage: the breach of duty was a direct cause and the proximate cause of the injury/damage (losses which may be pecuniary or emotional).

These are the basic elements that are essential to prove the case of negligence. Above elements were well discussed in Bolam v. Friern Hospital Management Committee (1957) and Jacob Mathew v. State of Punjab (2005). All the above criteria must be satisfied in cases of professional negligence. The standard of skill and care required of every healthcare provider in rendering professional services or health care to a patient shall be that degree of skill and care ordinarily employed in the same or similar field of medicine as the defendant, and the use of reasonable care and diligence. These issues have been discussed very well by the Supreme Court of India under Dr. Laxman Balakrishna Joshi v. Dr. Trimbak Bapu Godbole (1969).

All professionals are expected to exercise knowledge and skills which they profess to have beyond that of the non-professionals. However, this skill and knowledge is to be judged by criteria and standards determined by the profession itself. It is only when the skill and knowledge fall below the established standard is the practitioner guilty of being negligent. Further, the Supreme Court has defined this duty in the celebrity case Indian Medical Association v. Shantha (1997) as “In general, a professional man owes to his client a duty in tort as well as in contract to exercise reasonable care in giving advice or performing service”. On setting out parameters of the standard of care required, the Court held that the standard should be established by the profession.

Various acts falls within the ambit of medical negligence, as noted in a case K. Gracykutty v. Dr. Annamma Oommen and Another (1992), of the Kerala State Consumer Disputes Redressal Commission that the following acts would fall under the ambit of medical negligence:

1. Recklessness in undertaking a treatment or recklessness in the treatment of it
2. Indifference in handling of medical cases
3. Failure to act diligently and alertly at the appropriate time
4. Wrong diagnosis or treatment which under no norms of practice can be justified
5. Evident negligence like amputating a wrong limb or administering a prohibited or known counterproductive medicine
6. Misrepresenting that one possesses the skill or expertise which he/she does not possess.

Hence, all practitioners must bear in mind that they should bring in a reasonable degree of skill and knowledge, and must exercise a reasonable degree of care. Failure to exercise a reasonable degree of skill in diagnosis and providing care can constitute negligence.

13.1.1 The Tests of Liability: Bolam and Bolitho

The Bolam Test To succeed in proving liability, one needs to prove that the therapist was negligent and acted in a manner that no other therapist would have done. There are two tests of liability: the Bolam test and the Bolitho test.

The Bolam Test was developed by Bolam v. Friern Hospital Management Committee (1957), an English tort law case that develops rules for assessing standards of reasonable care in negligence involving professionals. Bolam was suffering from mental illness of the depressive type and was advised by the doctor attached to the defendants' Hospital to undergo electroconvulsive therapy. Prior to the treatment, Bolam signed a form of consent to the treatment but was not warned of the risk of fracture involved. Even though the risk was very small and on the first occasion when the treatment was given, Bolam did not sustain any fracture but when the treatment was repeated for the second time he sustained fractures. No relaxant drugs or manual control were used except that a male nurse stood on each side of the treatment couch throughout the treatment. About this treatment, there were two bodies of opinions, one of which favoured the use of relaxant drugs or manual control as a general practice, and the other opinion confined the use of relaxant drugs only to cases where there are particular reasons for their use and the Bolam case was not under that category. Ultimately the Court held the doctors were not negligent. In this context, the following principles were laid down: "A doctor is not guilty of negligence if he has acted in accordance with a practice accepted as proper by a responsible body of medical men skilled in that particular art". In other words this test states that if a therapist failed to reach the standard prescribed or practiced by a responsible body of psychotherapists' opinion, then it is termed as negligent.

The Bolitho Test The Bolitho test, developed by the Bolitho v. City and Hackney Health Authority (1993), modified the standard of care required by medical specialists in 'the Bolam Test'. In the case of Bolitho, the House of Lords would not accept a defence of being 'respectable', 'reasonable', or 'responsible' without first determining whether the professional opinion can withstand logical analysis. This Bolitho test further qualifies 'the Bolam test' with logical basis. This clearly says that the court had to be satisfied that the expert opinion had a logical basis.

In India, clients can approach Consumer Protection Act (CPA) 1986, which is a quasi-judicial mechanism set up at district, state, and national levels, where consumers can file their complaints, which are entertained by the judicial bodies referred to as consumer forums. The psychotherapy profession comes under the ambit of the CPA. As per the Consumer Protection Act, a consumer (client) is a person who hires or avails of any services for a consideration. Under this legislation, a therapist owes certain duties to the client, who consults him for his psychological problems. A deficiency in this duty results in negligence. A client can approach the consumer court, if he or she has suffered loss or damage as a result of any deficiency of services.

To summarize, professional negligence occurs when a therapist fails to act as a reasonable therapist would have acted under the circumstances. In the presented case vignette, Mr X committed suicide in spite of providing reasonable degree of care and if he had referred the case to another, which is endorsed by the professional body, then the therapist is not considered negligent but the unfortunate incident is an outcome of the disorder. A therapist must exercise that level of skill, diligence and judgment that a reasonable therapist would have exercised under the same or similar circumstances. On the other hand, if the therapist did not provide adequate reasonable degree of care then the therapist is liable for prosecution. The client can file a civil suit (for compensation) and a criminal case (for punishment), as well as in the Consumer Fora (to claim for damages), against the negligent psychotherapy practitioners. However, one should remember that psychotherapy for a consideration (paid) comes under the ambit of the Consumer Protection Act 1986. Free psychotherapy services do not come under the ambit of the legislation.

Case 13.2

Mrs. Y, 22 years old, a married engineer, from middle socio-economic status, diagnosed with schizophrenia, is on treatment from a government hospital. She is on medication and also on psychotherapy. She has already completed 12 sessions of supportive psychotherapy. During the sessions she had revealed certain private details about her premarital relationship with one of her college mates. She had requested for confidentiality. After a few weeks the therapist comes to know that Mrs.Y's husband has filed for divorce in the court of law and now he is asking for a copy of the medical record and also the psychotherapy notes.

What should the therapist do? Is the therapist legally obligated to keep the therapy session notes? Does the husband have a right to access the medical records of his wife?

13.2 Record keeping

Practitioners should be familiar with the legal and ethical requirements for record keeping in their specific professional contexts (see Case 13.2). A psychologist must make efforts to see that legible and accurate entries are made in client records as soon as is practicable after a service is rendered. The nature and extent of the record will vary depending upon the purpose, setting, and context of the psychological services (Bond and Mitchels 2014; Wheeler and Bertram 2008). The process of keeping records involves consideration of local legal requirements, ethical standards, supervisory needs, research protocol obligation as well as the demands of the particular professional context (Bond and Mitchels 2014; Luepker 2003). Appropriate records can also help protect both the client and the psychologist in the event of legal or ethical proceedings. Adequate records are generally a requirement for third-party reimbursement for psychological services.

Therapists are ethically and legally required to store records in a secure manner and to protect client confidentiality. Failing to maintain adequate client records potentially could lead to a malpractice claim because it breaches the standard of care expected of a mental health practitioner (Luepker 2012). The assurance of confidentiality is critical for the provision of many psychological services. Maintenance of confidentiality preserves the privacy of clients and promotes trust in the profession of psychology (American Psychological Association 2007).

Recordkeeping guidelines delineate three types of content (American Psychological Association 2007)

1. General file information: This includes demographic details, presenting problems, diagnosis, intervention plan, fee agreement, billing information and informed consent.
2. Documentation of service: This includes the date, duration and type of psychotherapy services and session notes. These notes should be updated for each contact with a client. Recognizing that client and other professionals may review these records (team members, colleagues and insurance agency), the practitioner may want to be sensitive to the language he or she uses to describe the patient.
3. Other information: A variety of other types of information may be included in the record, such as assessment data, crisis management documentation, consultation with other professionals, and telephone and email contacts.

American Psychological Association (2007) mandates psychologists to retain full records until seven years after the last date of service delivery for adults or until 3 years after a minor reaches the age of majority, whichever is later. If there is no state statute, the psychologist may be held to what is determined to be 'customary practice' in his or her area.

According to Medical Council of India (2002) guidelines

1. Inpatients records should be maintained in a standard proforma for three years from the commencement of treatment.
2. If any request is made for medical records either by the patients/authorized attendant or legal authorities involved, the same may be duly acknowledged and documents shall be issued within the period of 72 hours.
3. Maintain a register of certificates with the full details of medical certificates issued with at least one identification mark of the patient and his signature.
4. Efforts should be made to computerize medical records for quick retrieval.

13.2.1 Patient Access to Records

Patients have a right to access their personal health information both under Medical Council of India (MCI) Regulations (2002) and the Right to Information Act (2005). Certain personal records must be given to the patient as a matter of right. Discharge summary, referral notes, and death summary in case of natural death are important documents for the patient/immediate family member. These individual personal documents have to be given to the patients, even to those who leave against medical advice.

Physicians cannot refuse to grant a patient access to their records for the purpose of avoiding a legal proceeding. With the enforcement of the MCI Regulations (2002), it has been held without confusion that the patient has a right to claim medical records pertaining to his/her treatment and the hospitals are under an obligation to maintain them and provide them to the patient on request. Request for medical records by the patient or an authorized attendant should be acknowledged and the documents should be issued within 72 hours. The hospitals can charge a reasonable amount for the administrative purposes including photocopying the documents. Failure to provide medical records to patients on proper demand will amount to deficiency in service and negligence.

Certain records cannot be given to patients without the direction of the Court. The files of medicolegal cases including autopsy reports cannot be handed over to the patient or relatives without the direction of the Court. But if these medicolegal cases are being referred to another centre for management, copies of records could be given. Suggestions for patients general clinical record include:

1. Patient contact and basic information
2. Intake materials including the initial intake history, evaluation and assessment

3. Progress/session notes, initial and interval treatment/service plans and closing/termination summary
4. Referrals made, consultation reports and testing results obtained
5. Details of communication with other practitioners
6. Correspondence with the patient or non-clinical collateral contact details
7. Billing records
8. Informed consents for treatment, consents and authorizations to use and/or disclose clinical information/records.

The following information should be included when documenting a counselling session:

1. Subject being discussed;
2. Scope of the discussion (e.g. educational components, management options, prognosis);
3. Patient's response to the discussion;
4. Therapy prescribed (if any);
5. Action plan or goal including follow-up.

It is important to know that the information contained in the case belongs to the client. The client has the right, or privilege, that his or her information will be kept confidential. The case information in the client file should be considered a legal document that can be subpoenaed by the court.

Information recorded should be complete so anyone reading the notes can understand who this client is, what brought him/her to the therapist's office, the goals established, the clinical plan, interventions utilized, their effectiveness and plans for referral/follow-up, if any. A simple standard can be that documentation should enable anyone who reads notes to: understand the client's reasons for seeking treatment, treatment plan to address the presenting problem, results of the interventions employed and the client's disposition.

Documentation takes time and can be tedious. It is good for the practitioner to get into the habit of establishing some allocated time to get the documentation done. In many countries, therapists do an audio recording of the sessions for documentation purposes; however prior written consent of the patient needs to be taken during the intake session.

In general, a summary of psychotherapy is produced in case of legal matters but according to the current legal reforms it is required to have each sessions report otherwise it is amenable to liability. The new Mental Health Care Bill (Narayan and Shekhar 2013) mandates proper record keeping. The bill also dictates that the patient and the nominated representative have the rights to access the records. Against this background, an effort has been made to have uniform session reporting guidelines taking into consideration ethical and legal issues. These guidelines do not establish rules for practice, but rather provide an overall conceptual model and strategies for resolving divergent considerations.

To summarize, the records should reflect the therapist's competence, thoughtfulness, decision-making ability, and capacity to weigh available options, rational for treatment selection and knowledge of clinically, ethically and legally relevant matters (Simon and Sadoff 1992). In the Case 13.2 Mrs Y's psychotherapy session records and medical records cannot be given to her husband without Mrs. Y's consent. The records can be released to the court only after following the proper procedure. In this regard, the therapist can only release information relevant to the case, and the summoned records should be presented to the court in a sealed envelope marked 'confidential'. It is illegal to destroy or tamper with records to avoid disclosure.

Box 13.1: GUIDELINE ON SESSION REPORTING FORM

Name of the Patient:
Date: Time:
Session No:
Duration of the session:
Last session:
Review of the last session:
  Focus/Goal Planned: Short-term and Long-term goals
Emotional atmosphere during the session:
Main themes (Give details of the major issues discussed/explored/transpired):
Type of Psychotherapy approach used (Cognitive, Behavioural, Mindfulness, etc.):
Therapeutic interventions:
Comments on the session (Behaviour/Dynamics/Progress/Other Processes):
  Advice/suggestion given during the session:
Details of supervisory discussion (if any):
Next session on:
Plans for future interventions:
Medication (if any):

It is good practice to take the patient's consent before submitting the patient's record to the court. If the client wants the therapist to comply with the subpoena, then the therapist should get a written authorization from the client before sharing the requested information. If the client does not consent, the therapist can continue to assert the privilege, until a court hearing which may judge that the privilege should not be upheld. At this point, the therapist must abide by the court order (if not, a contempt-of-court citation can be issued) (Simon 2003; Simon and Sadoff 1992).

13.3 Confidentiality

Confidentiality is the secret-keeping duty that arises from the establishment of the professional relationship psychologists develop with their clients (Younggren and Harris 2008). This therapist–patient privilege only applies to secrets shared between the therapist and the patient during the course of providing care. This confidentiality dates back to at least the Hippocratic Oath, which reads

> Whatever, in connection with my professional service, or not in connection with it, I see or hear, in the life of men, which ought not to be spoken of abroad, I will not divulge, as reckoning that all such should be kept secret.

Traditionally, medical ethics has viewed the duty of confidentiality as a relatively non-negotiable tenet of medical practice. This is important for the therapeutic alliance, as it promotes an environment of trust. There are important exceptions to confidentiality, namely where it conflicts with the clinician's duty to warn or duty to protect. This includes instances of suicidal behaviour or homicidal plans, child abuse, elder abuse and dependent adult abuse (see Case 13.3).

Case 13.3
Mr. Z, a 32 years old unmarried engineer, from high socio-economic status, was diagnosed to be suffering from paranoid schizophrenia. He is on medication and also on psychotherapy. He has already completed three sessions of psychotherapy. During the fourth session, he reveals to the therapist that he is planning to harm Ms. X for refusing to marry him. Mr. Z has also revealed details of this plan to the therapist. Mr. Z has also requested his therapist not to reveal his plan to anyone.

What should the therapist do? Is therapist legally obligated to keep confidentiality? ÅDoes the therapist owe a duty to general public at large?

The Supreme Court of India has endorsed the principle of medical confidentiality as deriving from the classical conception of the right to privacy. Doctors are morally and ethically bound to maintain confidentiality. In such situations, public disclosure of even true private facts may amount to an invasion of the Right of Privacy which may sometimes lead to the clash of a person's 'right to be left alone' with another person's right to be informed.

13.3.1 Confidentiality and the Public Interest

Public interest requires that confidences be maintained. Breaking confidence is possible if the information is:

1. In the public interest (to prevent/report crime, malpractice, suicide, professional misconduct, suspected child abuse)
2. Related to illegality or breaking of law
3. Already in public domain

Legal opinion has supported the principle of the counsellor's 'fiduciary duty' of confidentiality in student counselling.

13.3.2 Confidentiality and Privilege

1. Confidentiality arises where trust is reasonably expected in a relationship.
2. Marital negotiations for reconciliation or separation can be privileged from disclosure.
3. Professional privilege (protection against compulsory disclosure of confidences) applies only to client–lawyer communications for legal advice.

The Supreme Court of United States created a psychotherapist–patient privilege in the Federal Rules of Evidence Jaffee v. Redmond (1996). The Court also noted that in a draft of the Federal Rules of Evidence, the Congress had listed *psychotherapist–patient privilege* as one of nine privileges, but in the end decided that the determination of privilege should be left to the courts. Hence, it clearly dictated that confidentiality is not absolute. A legitimate breach of confidentiality is upheld in reporting suspected child (sexual) abuse by their contract of employment or professional code of practice. Similar is the case of breaking of confidentiality in the public interest, in order to report serious crimes such as providing information to legal agencies regarding terrorist offences.

13.3.3 Child Abuse and Confidentiality

Therapists must make a report to an appropriate authority upon knowing or suspecting the occurrence of child abuse. There are no time limits on child abuse reporting in the sense that as long as the victim is still a minor, therapists have an obligation to file a child abuse report. Child sexual abuse laws in India have been enacted as part of the nation's child protection policies. The Parliament of India passed the 'Protection of Children against Sexual Offences Bill, 2011' (POCSO Act, 2012) regarding child sexual abuse on May 22, 2012 into the Act. When a doctor or a therapist has reason to suspect that a child has been or is being sexually abused, he/she is required to mandatorily report this to the appropriate authorities (i.e. the police or the relevant person within his/her organization who will then have to report it to the police). Failure to do this would result in imprisonment of up to 6 months, with or without fine (Section 21, Protection of Children from Sexual Offences Act, 2012).

Making a report to the police under Section 19(1) of the POCSO Act (2012) states that any person, including a member of a non-governmental organisation (NGO), may make a report under this section. Many NGOs work closely with vulnerable children and are in a position to detect child abuse. In many cases, a child may feel more comfortable disclosing abuse to an NGO worker rather than to someone from his/her own family. An NGO that has the knowledge of sexual abuse of a child is also bound by the principle of mandatory reporting under Section 21(1) of POCSO Act, 2012.

13.3.4 Suicidal or Homicidal Clients and Confidentiality

A breach of confidentiality is permitted when a client poses an imminent danger to himself/herself or to others. Failure of a therapist to ensure client safety within a high risk for suicide situation could end in harm or death to the client. Therefore, therapists must weigh consequences of breaking confidentiality versus potential client harm. Self-inflicted harm or damage to property is not a strong enough public interest area to infringe upon the need for confidentiality. Procedurally, the therapist's response to suicidality can include formulating a 'suicide prevention contract' with the client, informing the client's family, and/or having the client hospitalized. If possible, the therapist can discuss his/her intended action to resolve the situation with the client. Communication with others may be best limited to information pertinent to the present situation to protect client confidentiality.

In Tarasoff v. Regents of the University of California (Beck 1985) case, the California Supreme Court found that, despite patient–psychotherapist confidentiality, a duty to warn exists when the therapist determines a warning is essential to avert a danger rising from the patient's condition. The mental health profession quickly responded to this decision, claiming that they have no inherent ability to predict violence and that such a ruling violated their 'special' relationship and would prevent patients from trusting them. It could also generate false positive predictions as a means of diverting liability just in case something happened. Overall, this would be a detriment to those needing treatment, as well as a deterrent to clients who might otherwise expose their violent fantasies.

The court agreed to rehear the case and issue a second opinion. It still found that therapists have a duty to potential victims, but they need only use 'reasonable care' to protect the person. That is, the therapist needs to consider admission or hospitalize the patient to avoid the potential for harm.

To summarize, confidentiality is not absolute. In case of a homicidal or suicidal threat, the therapist has the responsibility to warn the potential victim or to inform the police regarding the risk, when the therapist determines a warning is essential to avert a danger rising from the patient's condition. At the same time, the therapist should consider admission of the violent patient in a mental hospital.

13.4 Boundary Crossing and Violation

A boundary is a line, point or plane that indicates or fixes a limit or extent. Boundaries are also the limits we set in relationships, which establish clear roles for therapists and define the therapeutic territory (Herlihy and Corey 2014; Lamb and Catanzaro 1998; Simon 1992). They do not undermine the therapist–patient relationship (Gabbard 1997; Smith and Fitzpatrick 1995). If boundaries are ignored, therapists can find themselves acting in their own best interest instead of the patient's best interest and may exploit the patient. Boundaries are derived from ethical treatise, cultural morality and jurisprudence. Sometimes, it is difficult to clearly define the perimeter of these boundaries and the integrity of the relationship (Gutheil and Simon 2002). As per researchers (Gutheil and Gabbard 1993), there are two types of boundary issues, (a) Boundary crossings and (b) Boundary violations. Boundary crossings are harmless deviations from traditional clinical practice, behaviour, or demeanour. Neither harm nor exploitation is involved. For example, giving a patient an emergency taxi fare in a heavy downpour, or accepting cake on a client's birthday, etc. Boundary violations, in contrast, are typically harmful and are usually exploitive of patients' needs—erotic, affiliative, financial, dependency or authority. Examples include, having sex with client or for financial demands beyond the fee, etc. (Folman 1991; Gabbard 1997; Gutheil and Gabbard 1993; Norris et al. 2014).

The American Counselling Association (1995), the American Association for Marriage and Family Therapy (AAMFT 2015, 1.4, 1.5), and the American Psychological Association (APA 2010, 10.5, 10.08) all agree that sexual contact before two years after therapy termination is unethical. All of these associations indicate that in the case of sexual relations with former clients, even after two years have elapsed, the burden of demonstrating the absence of exploitation rests with the therapist (see Case 13.4).

Case 13.4

Ms. M, 22 years, an unmarried engineer, from high socio-economic status, diagnosed with Borderline Personality Disorder. She is on medication and also on psychotherapy. She has already completed nine sessions of psychotherapy. During the psychotherapy sessions, Ms. M, gets attracted to her therapist. She starts giving gifts (such as sweets, flowers, greeting cards, watch, etc.) to her therapist. Therapist also gets involved romantically with his client and has a physical relationship with Ms. M, after consent from her. Now, they both are planning to get married.

Is it acceptable for a therapist to get into a consensual sexual relationship with his client? Can the therapist marry his client?

Following are some suggestions which are helpful when a boundary crossing causes, or seems to be leading towards, serious problems (Pope and Keith Spiegel 2008). Therapists need to continue to monitor the situation carefully. Be open and non-defensive. Talk over the situation with an experienced colleague who can provide honest feedback and thoughtful consultation. Try to see the matter from the client's point of view. Keep adequate, honest, and accurate records of this situation as it evolves. If you believe that you made a mistake, consider apologizing. Sometimes if the emotional turmoil is too much, termination of therapy is advised, with referral of the client to another therapist (Pope and Keith Spiegel 2008).

To summarize, psychotherapists need to be aware of their influential position with respect to their clients and need to avoid sexual contact with their clients, whether such contact is consensual or forced. It is unethical and also illegal.

13.5 Practical Issues

In case of personal problems and conflicts, psychotherapists should recognize that their personal problems and conflicts can interfere with the effectiveness of their work (Pope and Keith Spiegel 2008). In such cases, they should refrain from that particular work (therapy, research) and seek help for the same.

Late to the session It is not uncommon that clients report late to the session. Though it is difficult for the therapist to conduct session in such instances owing to time constraints, it is advisable to discuss the difficulties in reporting on time and the possible solutions for the same.

Cancellation Generally, clients are asked to inform therapist at least 24 hours in advance if they want to cancel a session. Though there is a practice of charging clients if they fail to do so, it is rarely done in the Indian setting.

Emergency contact Clients/family may be provided with emergency contact numbers of the emergency psychiatric settings so that they do not have to wait in distress till the designated appointment date for session. Client and family members also have to be made aware that therapist's privacy and personal time needs to be respected. If a client is making repeated phone calls it would be prudent to explain the client about the professional relationship and request to call only during emergencies. If the client continues to make phone calls it would be advisable to block the number during non-working hours. Hence, it is advisable that all therapists should avoid giving personal mobile number, residential phone number and residential address. If a patient threatens self-harm over the phone, the therapist should contact family members of the client, explain the situation and make a referral to the nearest emergency mental health centre or hospital. If no family members are available or traceable, it would be prudent to inform the law enforcing agencies about the client's self injurious behaviour to protect him/her.

Interruption of services Psychotherapists have to take steps for the provision of uninterrupted services to the client in circumstances such as psychotherapist's illness, death and therapist's or client's relocation. Appropriate referral has to be made when client has financial constraints to pay for the services. In this regard, abrupt ending of sessions is not appropriate and ethical (American Psychological Association 2002).

Termination Psychotherapy is terminated when the therapist evaluates that the client does not need therapy any more, is not benefitting from the same or is counterproductive for the client (American Psychological Association 2002). Termination of sessions needs to be planned and conveyed to the client well in advance. However, severe pathological transference may call for abrupt termination of the sessions with proper documentation of the reason for termination and referral of the client to another therapist. This is also true in case of severe counter-transference, during which the therapist should seek consultation.

Gifts Currently, the empirical data available on the impact of gift-giving by therapist or client is sparse. In one study (Knox et al. 2003) therapist's experiences of receiving gifts from clients suggested that gift-giving had symbolic value which ranged from appreciation, boundary violation to manipulation. Therapists needs to be aware of the ethical guidelines, cultural norms related to gift-giving or accepting gifts. There is need for clear communication to clients at the outset of therapy if the therapist has a strict 'no-gifts-policy' as well as discussions about the client's emotions and experience in the context of therapist refusal to accept a gift.

Self–disclosure In a qualitative study (Audet and Everall 2010), it was found that therapist self-disclosure has both facilitating (closeness, comfort, feeling understood/not judged) and hindering effects (role confusion, role reversal, feeling misunderstood, etc.). Some authors support a balanced and careful use of self-disclosure without indulging in disclosing too much or too little (Bloomgarden and Mennuti 2010). Newer ethical dilemmas are emerging with respect to client-therapist personal contact via social media.

Telephone counselling Therapists need to avoid telephone counselling in regular counselling services. Telephone counselling needs to be used only during crisis. Ethical and legal norms have not been well established in telephone counselling.

To conclude, psychotherapist–client relationship is a complex and multifaceted. In the era of consumerism, the therapist needs to be aware of the client's rights, ethical issues and prevailing legal system. Professional codes of ethics are fundamental for ethical practice; however simply knowing these codes is just the beginning. The ability to think critically and apply general ethical principles to specific situations is vital. Relevant laws vary substantially from state to state and from context to context and from time to time. When appropriate, practitioners are encouraged to consult legal counsel who can review the pertinent law and facts and provide legal assistance as needed.

References

American Association for Marriage and Family Therapy. (2015). AAMFT code of ethics. Retrieved from http://www.aamft.org/iMIS15/AAMFT/content/legal_ethics/code_of_ethics.aspx.

American Counselling Association. (1995). *Code of ethics and Standards of Practice*. Alexandria: Author.

American Psychological Association. (2007). Record Keeping Guidelines. Retrieved from http://www.apa.org/practice/guidelines/record-keeping.aspx.

American Psychological Association. (2002). Ethical principles of psychologists and code of conduct. *American Psychologist, 57*(12), 1060–1073.

American Psychological Association. (2010). Ethical principles of psychologists and code of conduct. Retrieved from https://www.apa.org/ethics/code/.

Audet, C. T., & Everall, R. D. (2010). Therapist self-disclosure and the therapeutic relationship: A phenomenological study from the client perspective. *British Journal of Guidance & Counselling, 38*(3), 327–342.

Austin, K. M., Moline, M. E., & Williams, G. T. (1990). *Confronting malpractice: Legal and ethical dilemmas in psychotherapy*. Thousand Oaks, California: Sage Publications, Inc.

Beck, J. C. (1985). *The potentially violent patient and the Tarasoff decision in psychiatric practice*: American Psychiatric Press.

Bernstein, B. E., & Hartsell Jr, T. L. (2004). *The portable lawyer for mental health professionals: An AZ guide to protecting your clients, your practice, and yourself*: Wiley.

Bloomgarden, A., & Mennuti, R. B. (2010). *Psychotherapist revealed: Therapists speak about self-disclosure in psychotherapy*. New York: Routledge.

Bolam v. Friern Hospital Management Committee. (1957) 1 WLR 582.

Bolitho v. City and Hackney Health Authority. (1993) P.I.Q.R P334.

Bond, T., & Mitchels, B. (2014). *Confidentiality & Record Keeping in Counselling & Psychotherapy*: SAGE.

Croxton, T. A. (1985). The therapeutic contract. In S. Martin, G. Paul, S. Rosemary & V. Robert (Eds.), *Individual change through small groups* (pp.159–179). New York: The Free Press.

Dr. Laxman Balakrishna Joshi v. Dr. Trimbak Bapu Godbole. (1969). AIR 1969 SC 128.

Feldman, S. R., & Ward, T. M. (1979). Psychotherapeutic injury: reshaping the implied contract as an alternative to malpractice. *North Carolina Law Review, 58*, 63.

Folman, R. Z. (1991). Therapist-patient sex: Attraction and boundary problems. *Psychotherapy: Theory, Research, Practice. Training, 28*(1), 168.

Gabbard, G. O. (1997). Lessons to be learned from the study of sexual boundary violations. *Australasian Psychiatry, 31*(3), 321–327.

Gutheil, T. G., & Gabbard, G. O. (1993). The concept of boundaries in clinical practice: theoretical and risk-management dimensions. *The American Journal of Psychiatry, 150*(2), 188–196.

Gutheil, T. G., & Simon, R. I. (2002). Non-sexual boundary crossings and boundary violations: the ethical dimension. *Psychiatric Clinics of North America, 25*(3), A585–A592.

Herlihy, B., & Corey, G. (2014). *Boundary issues in counseling: Multiple roles and responsibilities*: Wiley.

Indian Medical Association v. V.P.Shantha. 1997(1) CPR377.

Jacob Mathew v. State of Punjab. (2005) 6 SCC 1.

Jaffee v. Redmond. (1996) 518 U.S.

Jenkins, P. (2002). *Legal issues in counselling & psychotherapy*: Sage.

K. Gracykutty v. Dr. Annamma Oommen and Another. 1992(1) CPR256.

Knox, S., Hess, S. A., Williams, E. N., & Hill, C. E. (2003). " Here's a little something for you": How therapists respond to client gifts. *Journal of Counseling Psychology, 50*(2), 199.

Lamb, D. H., & Catanzaro, S. J. (1998). Sexual and nonsexual boundary violations involving psychologists, clients, supervisees, and students: Implications for professional practice. *Professional Psychology: Research and Practice, 29*(5), 498.
Luepker, E. T. (2003). *Record keeping in psychotherapy and counseling*. New York: Brunner-Routledge.
Luepker, E. T. (2012). *Record keeping in psychotherapy and counseling: Protecting confidentiality and the professional relationship*: Routledge.
Medical Council of India. (2002). Code of Ethics regulations. Retrieved from http://www.mciindia.org/RulesandRegulations/CodeofMedicalEthicsregulations2002.aspx.
Ministry of Law and Justice. (2005). *Right to Information Act*. Retrieved from http://righttoinformation.gov.in/rti-act.pdf.
Ministry of Women and Child Development. (2012). The Protection of Children against Sexual Offences Bill. Retrieved from http://wcd.nic.in/sites/default/files/childprotection31072012.pdf.
Narayan, C.L., & Shekhar, S. (2015). The mental health care bill 2013: A critical appraisal. *Indian Journal of Psychological Medicine*, *37*(2), 215–219.
National Consumer Disputes Redressal Commission. (1986). *Consumer Protection Act*. Retrieved from http://www.ncdrc.nic.in/1_1.html.
Norcross, J. C., VandenBos, G. R., & Freedheim, D. K. (2011). *History of psychotherapy: Continuity and change*: American Psychological Association.
Norris, D. M., Gutheil, T. G., & Strasburger, L. H. (2014). This couldn't happen to me: boundary problems and sexual misconduct in the psychotherapy relationship. *Psychiatric Services, 54*(4), 517–522.
Packman, W. L., O'Connor Pennuto, T., Bongar, B., & Orthwein, J. (2004). Legal issues of professional negligence in suicide cases. *Behavioral Sciences & the Law, 22*(5), 697–713.
Pope, K. S., & Keith Spiegel, P. (2008). A practical approach to boundaries in psychotherapy: Making decisions, bypassing blunders, and mending fences. *Journal of Clinical Psychology, 64*(5), 638–652.
Schutz, B. M. (1982). *Legal Liability in Psychotherapy: A Practitioner's Guide to Risk Mangement*: Jossey-Bass.
Sills, C. (2006). *Contracts in Counselling & Psychotherapy*: Pine Forge Press.
Simon, R. (1991). The practice of psychotherapy: Legal liabilities of an "impossible" profession. *American Psychiatric Press Review of Clinical Psychiatry and the Law, 2*, 3–91.
Simon, R. I. (1992). Treatment boundary violations: Clinical, ethical, and legal considerations. *Journal of the American Academy of Psychiatry and the Law Online, 20*(3), 269–288.
Simon, R. I. (2003). *Clinical psychiatry and the law*: American Psychiatric Publication.
Simon, R. I., & Sadoff, R. L. (1992). *Psychiatric malpractice: Cases and comments for clinicians*: American Psychiatric Publication.
Smith, D., & Fitzpatrick, M. (1995). Patient-therapist boundary issues: an integrative review of theory and research. *Professional Psychology: Research and Practice, 26*(5), 499.
Sulzer, E. S. (1962). Reinforcement and the therapeutic contract. *Journal of Counseling Psychology, 9*(3), 271.
Thompson, A. (1983). *Ethical concerns in psychotherapy and their legal ramifications*. Lanham: University Press of America.
Van Hoose, W. H., & Kottler, J. A. (1985). *Ethical and legal issues in counseling and psychotherapy*: Jossey-Bass.
Wheeler, A., & Bertram, B. (2008). *The counselor and the law*. Alexandria: American Counseling Association.

Wolberg, L. R. (1951). Supervision of the psychotherapeutic process. *American Journal of Psychotherapy, 5*, 147–171.

Wolberg, L. R. (1988). *The technique of psychotherapy, Parts 1 & 2: Grune & Stratton*. Bra: Inc/Harcourt.

Younggren, J. N., & Harris, E. A. (2008). Can you keep a secret? Confidentiality in psychotherapy. *Journal of Clinical Psychology, 64*(5), 589–600.

Chapter 14
Research Ethics in Psychotherapy and Psychosocial Interventions: Role of Institutional Ethical Review Boards

Shoba Srinath and Poornima Bhola

Abstract Research on psychosocial interventions looks at answering questions about the effectiveness of counselling and psychotherapy and also attempts to uncover the crucial ingredients and change processes. Maintaining the ethical frame is paramount in the therapy room and this extends to any intervention research which involves human participants. The counselling or psychotherapy research context has some unique characteristics; there is a power dynamic with the psychologically distressed patient or client in a position of relative vulnerability vis-à-vis the therapist–researcher. The content and processes of interactions in psychosocial intervention give rise to complexities that are different from those in drug trials. The chapter examines the debates and recommendations concerning key ethical issues in research on psychotherapy and counselling. These include informed consent, disclosure and deception, comparison groups, compensation and remuneration for research participation, research with vulnerable participants and issues, therapist competence and treatment integrity, intellectual property, copyright and publication ethics. Unique ethical questions may arise in multi-site and international collaborative research, particularly involving resource-poor countries like India. Working with vulnerable participants and issues, for instance refugees, children and persons with psychiatric disorders or trauma, increases the ethical responsibility of the researcher. Emergent ethical concerns stem from the advent of information technology and research explorations in the digital space. The critical role of institutional ethical review boards in evaluating and monitoring research proposals, educating researchers and supporting a climate of public accountability, is discussed.

S. Srinath (✉)
Department of Child and Adolescent Psychiatry,
National Institute of Mental Health and Neuro Sciences,
Bengaluru 560029, India
e-mail: shobasrinath@gmail.com

P. Bhola
Department of Clinical Psychology, National Institute of Mental Health
and Neuro Sciences, Bengaluru 560029, India
e-mail: poornimabhola@gmail.com

P. Bhola and A. Raguram (eds.), *Ethical Issues in Counselling and Psychotherapy Practice*, DOI 10.1007/978-981-10-1808-4_14

Keywords Research ethics · Informed consent · Institutional ethical review boards · Psychotherapy research

14.1 Balancing Scientific and Ethical Merit in Research

Research on psychosocial interventions looks at answering questions about the effectiveness of counselling and psychotherapy and also attempts to uncover the crucial ingredients and change processes. Maintaining the ethical frame is paramount in the therapy room and this extends to any intervention research which involves human participants (Garfield 1987). The Declaration of Helsinki (World Medical Association 2001) states that "considerations related to the wellbeing of the human subject should take precedence over the interests of science and society". The two basic principles of beneficence (achieving the greatest good) and non-maleficence (causing the least harm) form the ethical lens through which research proposals are viewed. The risk-benefit dialectic must be recognized and evaluated, with an understanding of the range of possible risks to participants; inconvenience, physical risk, psychological or emotional risk, social risk, economic or legal risk (Sieber 2001).

In the current climate of evidence-based interventions, there is a growing focus on research in order to establish which interventions work, for whom and under which conditions. The counselling or psychotherapy research context has some unique characteristics; there is a power dynamic with the psychologically distressed patient or client in a position of relative vulnerability vis-à-vis the therapist–researcher. This could be further heightened in socio-cultural contexts like India, with issues like poverty and the hierarchical 'doctor-patient' relationship (Wig 2004). The content and processes of interactions in psychosocial intervention give rise to complexities that are different from those in drug trials. Negotiating the dual roles of therapist and researcher can be challenging and may result in additional ethical dilemmas (Etherington 1996; Hart and Crawford-Wright 1999). The practice-research gap is narrowing and patient assessments may often be integrated into service delivery systems (Kass et al. 2013).

The most recent version of the Declaration of Helsinki (World Medical Association 2013) provides detailed guidelines for biomedical research and these are broadly applicable to research on psychosocial interventions as well. Many professional organizations for psychiatrists, psychologists, counsellors and psychiatric social workers have ethical guidelines which include a focus on research ethics (Agarwal 2010; American Psychological Association 2010; Indian Association of Clinical Psychologists 2015; Indian Council of Medical Research 2006) and some guidelines address issues specific to researching counselling and psychotherapy (Bond 2004). Institutional Ethical Review Boards are an important resource for an independent review of research proposals to ensure the ethical conduct of research and protect the rights of participants. While these standards and regulatory mechanisms play an important role, an attitude of ethical mindfulness must be cultivated among researchers.

The chapter examines the debates and recommendations concerning key ethical issues in research on psychotherapy and counselling. These include informed consent, disclosure and deception, comparison groups, compensation and remuneration for research participation, research with vulnerable participants and issues, therapist competence and treatment integrity, intellectual property, copyright and publication ethics, multi-site and international collaborative research, research with vulnerable participants and issues and research ethics associated with information technology and the Internet. The critical role of institutional ethical review boards in evaluating and monitoring research proposals, educating researchers and supporting a climate of public accountability, is discussed.

14.2 Informed Consent

Informed consent is the cornerstone of ethically guided research and all potential participants must have adequate information about the research and make a voluntary decision about their involvement (Gupta & Kharawala 2012). Chaturvedi and Somashekar (2009) discuss the additional ramifications to informed consent when working with persons with psychiatric disorders. Some considerations include the evaluation of the competence of persons with psychiatric disorders to provide informed consent, concerns about the validity of consent in persons with impaired judgment or insight and the process of proxy consent provided by a patient's relative or caregiver. There are concerns regarding covert pressures for consent (Fulford and Howse 1993) and researchers are responsible for ensuring an ethical process for seeking and obtaining agreement for participation in studies.

A model to assess competence to consent has been proposed by Grisso and Appelbaum (1998) and includes four abilities. (i) The ability to express a choice (ii) The ability to understand information relevant to decision-making (iii) The ability to appreciate the significance of that information for one's own situation, especially concerning one's illness and the probable consequences of one's treatment options; and (iv) The ability to reason with the relevant information so as to engage in a logical process of weighing treatment options.

Competence to provide informed consent may vary across diagnostic groups. In a study by Gandhi and Vankar (2007), 64 % of psychotic inpatients exhibited inadequate treatment-related decision-making capacity and this was associated with low education, lack of employment, high level of psychopathology and cognitive dysfunction. A review by Hindmarch et al. (2013) suggested that depression can impair decision-making capacity reviewed by impacting on the ability to *appreciate* the nature and significance of their decision, and to a lesser extent on their understanding and reasoning ability.

Questions about the competence of psychiatric patients have consistently emerged and there are suggestions that structured processes or instruments can aid in the evaluation of decision-making capacity. Appelbaum et al. (1999) used the MacArthur Competence Assessment Tool-Clinical Research to assess the competence of depressed inpatients to consent to participation in a psychotherapy treatment project. While the overall decision-making capacity was adequate, some subjects appeared confused about the extent to which decisions about assignment to treatment groups would be made on the basis of their clinical condition rather than randomly. There was no correlation between severity of depression and the decision-making capacity. Chaturvedi (2008) highlights the need to develop locally relevant and culturally sensitive structured assessment tools to assess capacity to consent and the formulation of appropriate guidelines for seeking informed consent.

Should the consent process be video-recorded, particularly in situations where there are concerns about capacity or the potential participants are drawn from a vulnerable group? The procedure has recently been recommended for use in clinical trials. Chauhan et al. (2015) discussed some challenges related to confidentiality and the sociocultural environment in India, which posed barriers in implementing this guideline. They reported that 34 % of study participants in a rural community in South India refused consent for the audio-visual recording of the consent process. This trend was more marked among women and young adults.

Some of the key components of an informed consent form are outlined.

Components of an Informed Consent Form

An informed consent form should contain clearly worded information written for lay persons, in a language understood by the participant.

- The nature and purpose of the study, stating it as research
- State why the participant has been chosen for this research
- The duration of participation, type of intervention and procedures to be followed (including any investigations or samples, if any)
- Foreseeable risks and discomforts adequately described and whether the participation in the study involves more than minimal risk
- Potential benefits to participant, community or the profession, as may be applicable
- Voluntary participation and no loss of any benefits on withdrawal from the study at any point. Information about alternatives to participation, if relevant
- Steps taken for ensuring confidentiality and any limits to confidentiality
- Policy on any compensation for travel or any other remuneration for participation
- Information on data sharing if relevant, specifying how data will be shared and whether data will be anonymous
- Contact details of the Principal Investigator (PI) or local PI/Co-PI in multicentre studies for asking more information related to the research or in case of any concerns
- Contact details of the Institutional Ethical Review Board which approved the research protocol

- Signature of investigators and signature of participants indicating consent. For illiterate participants, a thumbprint and signature of a literate witness is typically required.

The World Health Organisation website has a series of informed consent form templates for clinical studies, qualitative studies and research involving minors/children. This useful online resource (http://www.who.int/rpc/research_ethics/informed_consent/en/) provides examples of key questions that can be asked at the end of each section to assess the participant's understanding of the information provided.

The consent process is different when the participants are minors or have limited capacity. Typically, assent is sought from the minor with a form worded at an appropriate language level. Assent refers to willingness to participate in research by a minor who cannot provide informed consent but is old enough to understand the proposed research in general, its expected risks and possible benefits, and the activities expected as a participant. Written informed consent is also sought from the parent/guardian. This is a debatable issue with some concerns raised about the minor's capacity to assent, the need to use an age threshold versus an individualized approach to assess capacity (Waligora et al. 2014) and the level of parental duress that might compromise the minor's autonomy in the decision-making process. Different countries or states within countries may have different definitions of the legal age of consent and researchers must be aware of these variations. If the minor reaches legal age of consent, another opportunity must be given to provide informed consent before continuing in the study.

Informed consent is best seen as an ongoing process (West 2002) and would need to be renewed if the research protocol is amended in any way or during the course of a longitudinal study.

14.3 Disclosure and Deception in Research

Is full disclosure to participants about a research study necessary, practical and desirable? (Fulford and Howse 1993; Lindsey 1984). The therapeutic relationship is built on a foundation of fidelity and this can complicate decisions about the degree of disclosure in psychotherapy research. Lindsey (1984) asserted that providing complete information on a consent form could make it long and complicated, and make it less likely that the form is read or understood. In addition, telling a potential participant 'everything' is difficult to achieve. This might also be ethically undesirable, seeming to put the onus entirely on the participant while absolving the researcher of responsibility in the process (Wing 1981). The mandate is to provide 'reasonable' information to enable an informed decision-making process.

A contentious issue in research ethics concerns studies that have an element of deception where the true purpose of the study is withheld or distorted. The deception is usually proposed because divulging the true nature of the study would alter participants' behaviours in a way that makes the study untenable. In these rare circumstances, 'debriefing', where a post hoc explanation is provided to participants, is an ethical obligation. The American Psychological Association guidelines (American Psychological Association 2010) assert that this debriefing must be conducted as early as feasible as and not later than the end of data collection, with participants given the option to withdraw their data. Deception is not permissible for any study that is likely to cause significant physical pain or emotional distress. Deception violates the principles of informed consent, and in a profession that emphasizes trust in the relationship, must be used only when the risks are minimal and are significantly outweighed by the potential benefits of the research (Lindsey 1984).

Some 'deceptive practices' could be ethically defensible in a research study as illustrated in a study of cognitive-behavioural intervention to reduce alcohol use among HIV infected outpatients in Western Kenya (Papas et al. 2012). Women participants, in particular, opted for text message reminders for sessions, such as "meet you at the shop tomorrow." While this practice was somewhat deceptive, the researchers rationalized that this served to protect the clients who were apprehensive of the consequences of disclosure within their socio-cultural realities. Most importantly, this not a unilateral researcher-led strategy but a collaborative decision involving the participants in the research.

14.4 Comparison Groups

Studies that examine the processes and outcomes of therapeutic interventions often have control groups built into the research design. O'Leary and Borkovec (1978) detailed how a psychotherapy placebo group (i.e. a theoretically inert procedure without specific activity for the condition being treated) may be unethical, impractical or methodologically problematic in psychotherapy research. Instead they suggested that alternative terms, such as 'best available' therapies, be used.

One of the important questions that this research attempts to answer is whether the therapeutic intervention is more effective than the usual community standards of care (TAU—treatment as usual). From an ethical standpoint, the selection of TAU as the comparison group is not always a straightforward and acceptable choice (Arean and Alvidrez 2002). In some areas there is no standard treatment available or the TAU could in itself be an unethical or proven ineffective option.

Weijer et al. (2011) describe the complexities of research ethics in a cluster randomized trial of group interpersonal psychotherapy for depression in rural Uganda (Bolton et al. 2003). They discuss the ethical issues that arise due to both the cultural differences and also the poor access to mental health care in developing nations. For example, what are the researcher obligations to provide therapeutic intervention for participants in the control group in a resource-poor country, who have no access to any 'treatment-as-usual' option? The Ottawa Ethics of Cluster Randomized Clinical Trials Consensus Group (Weijer et al. 2012) debated on the provision of augmented care for the control arm in such situations. They recognized that this could interfere with the scientific validity of the study and suggested that researchers consider the relative merits and demerits of this approach. Another suggestion is to ensure that the TAU control group receives the treatment at a later stage. Binik et al. (2011) raise concerns that this may pose practical, conceptual or ethical difficulties.

The concept of 'clinical equipoise', suggested by Freedman (1987) continues to guide the selection of comparative treatment arms in a trial. There should be genuine uncertainty within the professional community about which is the superior treatment. The research design cannot include a comparison group which is known to be less effective. Clinical equipoise is what justifies the random assignment of participants to treatment groups in randomized controlled trials. Although clinical equipoise is a necessary condition for random assignment, it may not be a sufficient one (Binik et al. 2011). There are diverse arguments and interpretations of the concept of clinical equipoise in selecting comparison groups and Binik et al. (2011) present a balanced perspective.

An additional ethical consideration is the need to ensure that participants have adequate information about how they will be assigned to the different treatment options. The ongoing monitoring of research trials is essential to look out for adverse effects and relative benefits of the treatment options. More often discussed in studies that involve medication treatment arms, is the possible halting of the trial and 'unblinding' of participants (if relevant), in the event of clear evidence of adverse effects or significant advantage of one treatment option over the other.

Contemporary psychotherapy research is influenced by the healthcare environment in which it is conducted and this also impacts on the choice of control groups. Ethical viewpoints need to be considered against the backdrop of scientific and practical considerations. In this context, Freedland et al. (2011) discuss the variants of treatment as usual control groups that may be used and also acknowledge the ethical tensions and issues associated with different choices.

14.5 Compensation and Incentives for Participation in Research

The decision to participate in research needs to be free from any overt or covert coercion or inducement and must not be influenced by the participants' vulnerabilities or circumstances. At the same time, one is aware that participants, in either an experimental or a control group, may incur a loss of revenue during the study period. Is it ethical then, for research participants to shoulder these costs or do researchers have some ethical obligation?

Wendler et al. (2002) describe four types of payments related to research participation and illustrate the need to differentiate between them. These include (i) Reimbursements for expenses (e.g. bus fare), (ii) Compensation for the time and inconvenience involved in research participation (iii) Appreciation payments at token levels (e.g. gift certificates of small monetary value) and (iv) Incentive payments that offer amounts that are not limited to reimbursement, compensation or token levels.

Any remuneration offered to participants is scrutinized carefully for possible ethical implications. The American Psychological Association Ethics code (2010) advocates foresight and caution before offering any money or professional services in return for participation in a study. The Indian Council of Medical Research (2006) has clear cut guidelines for appropriate compensation, including provisions for free ancillary services and parameters for compensating those who withdraw from the study for different reasons.

Researchers need to approach this issue carefully, particularly in cases or contexts where there is concern about consent being influenced by payment. Good practice would include establishing a clear and explicit justification for paying participants to the ethics committee. This should include explicit guidelines for when and how the payment is made and a policy to ensure that participants who choose to withdraw from the research are not denied their due reimbursement, compensation or incentive (Wendler et al. 2002). At a larger level, institutional and organizational guidelines and prevalent cultural and societal norms influence decisions about offering reimbursements, compensation, incentives or other payments. Financial remuneration is also seen as facilitative for both researchers and participants but continues to be an ethically controversial topic (Permuth-Way and Borenstein 2009).

14.6 Research with Vulnerable Participants and Issues

Research endeavours exploring the impact of psychosocial stress on individuals often touch upon areas of vulnerability in people's lives. Such studies may include vulnerable groups, for example, minorities, institutionalized persons, prisoners, persons with serious medical or psychiatric conditions, children. People may be socially vulnerable due to age, disability, physical or mental health, social, relational or financial circumstances (Bond 2004). Research that explores sensitive issues, for example personal traumatic experiences, may serve to increase vulnerability in participants. The complexities increase when there are multiple intersecting vulnerabilities; e.g. child refugees.

Does this mean that researchers should protect themselves by excluding vulnerable groups or areas of study? While this exclusionary attitude is not recommended, researchers must be careful that vulnerable groups are not 'over-researched', or studied without a clear rationale where the study would well be done with a non-vulnerable group (Bond 2004). The Helsinki guidelines (World Medical

Association 2013) state that research with a vulnerable group is justifiable when it is responsive to the needs or priorities of this group and the research cannot be carried out in a non-vulnerable group. A corollary is that the group must incur some benefits from the knowledge, practices or interventions stemming from the research. Ford et al. (2009) assert that a categorical exclusion of certain vulnerable groups from research altogether may even increase their vulnerability by preventing the design and improvement of interventions tailored to their needs and this approach may violate principles of social justice.

Researchers must be aware of the intersections between sources of vulnerability and the design and implementation of research studies. Working with different vulnerable groups may be associated with varied challenges and researchers need to be cognisant of the risks and specific modifications required.

An illustrative example addresses ethical issues involving research in conflict settings (Ford et al. 2009). The authors present a thoughtful discussion of possible ethically questionable shortcuts employed when trying to collect high quality data in a conflict zone and the difficulties in balancing the role of a researcher vis-à-vis the need to provide humanitarian aid. Issues of confidentiality, informed consent and anonymity are heightened and researchers need to ensure that the step of obtaining informed consent is not compromised or eliminated. Interviewing vulnerable sub groups; e.g. recent victims of torture or extreme violence, should be considered with caution and the use of secondary data sources could be considered. Ethical dilemmas extend to the accurate dissemination of results in the face of political pressures in the arena of international politics. Clearly, research in these settings need to be informed not only by ethical review boards but also by international human rights and humanitarian laws.

The experiences of those who have grappled with and addressed ethical issues related to working with *specific* vulnerable groups can inform other researchers. For example, there are certain contentious ethical issues in research involving persons with HIV positive status (Muthuswamy 2005) and yet other ethical questions unique to research on trafficked women working in the sex industry (Cwikel and Hoban 2005).

There is a greater likelihood that research involving psychosocial interventions addresses sensitive issues and personal experiences. Uncertainty about the ethics of asking about experiences like abuse or interpersonal violence and the lack of adequate research on the cost-benefit balance means that there is much more room for a researcher's individual belief systems to influence these decisions. Becker-Blease and Freyd (2006) argue that an overly cautious stance ignores the possible costs of limiting research in these areas and also may overemphasize vulnerability among abuse survivors.

Researchers must factor in specific risks and power imbalances when working with specific vulnerable groups during the study design stage. In some cases, the consent form could mention the possibility of experiencing distress as a result of participation and provide resources for additional support and assistance.

14.7 Therapist Competence and Treatment Integrity

Psychotherapy is a relational enterprise and interventions are delivered by therapists. How can one draw conclusions about the processes, efficacy or effectiveness of psychosocial interventions if they are not delivered in a standard and competent manner? From an ethical standpoint, research participants must receive the best possible care from well-trained therapist-researchers. This concept of *treatment integrity* or the broader term of *therapy quality* (Fairburn and Cooper 2011) includes aspects of treatment *adherence* (whether the right therapeutic procedures were used) and *competence* (how well the chosen procedures were implemented). If therapist-researchers will be supervised to ensure adequate delivery of the intervention, this must be mentioned in both the research protocol and the informed consent form.

This has implications for training and supervision of therapist-researchers, clear operational definitions of treatment techniques and processes and the inclusion of measures of treatment integrity in research designs (Bond 2004).

14.8 Ethical Issues in Qualitative Research

Qualitative research designs operate within the generic framework of research ethics but these modes of inquiry may also give rise to unique ethical considerations (McLeod 2001; Richards and Schwartz (2002).

During qualitative studies, there may be some changes in the processes and intended use of the data as the research progresses (Richards and Schwartz 2002). Instead of taking a general informed consent only at the outset, Grafanaki (1996) proposed that ongoing renewed consent may be particularly relevant in some qualitative studies on counselling and psychotherapy. This process would serve to protect participants, enhance their freedom of choice and allow for the possibility of changes at different points in the research process. Despite possible practical difficulties related to the cost and effort of obtaining informed consent at different stages, Richards and Schwartz (2002) still advocate treating informed consent as a process.

Confidentiality requires special considerations in qualitative studies where personal narratives and information are collected. Specific strategies have been outlined for the secure storage of any tapes or transcripts, the use of pseudonyms or initials should be used in transcripts and alteration of any other identifying details. Realistically, it may be difficult to completely anonymize interview data at the stage of analysis and transcription and even careful use of protocols of anonymization could leave enough information for participants to be identified by themselves or others (Richards and Schwartz 2002). The informed consent form must clearly mention who has access to the data, how and where it will be stored, how long it will be stored and how the tapes or transcripts will be destroyed.

The inherent power imbalance between the researcher and participant is a concern across both quantitative and qualitative research designs. Where this is particularly salient in qualitative approaches to collecting data is when a researcher could be overstep boundaries while exploring sensitive issues and the participants may reveal more information that they anticipated during the consent process (Richards and Schwartz 2002). The vulnerability to blurring of boundaries is impacted by the degree of the power imbalance, length of contact between researcher and participant and the degree of clarity about termination of the interaction (Bourdeau 2000). The other caution is for the researcher to remain within the boundaries of their role and guard against dual relationships with participants; for instance as a therapist, advocate or supportive friend.

When qualitative studies explore sensitive issues or include in-depth interviews, there is the possibility that the process evokes negative emotions in participants. McLeod (2001) asks questions about the ethical obligations of a researcher when the participant is distressed following a detailed interview on a sensitive topic. Such concerns need to be anticipated during the planning stage and adequate debriefing methods and support pathways must be provided. Equally, the researcher could also experience difficult emotions arising from the interaction. Adequate preparation, debriefing, support and consultation avenues are recommended for the researcher (Wiles 2013).

Compared with quantitative data, the interpretations of qualitative data tend to be more subjective and may vary depending on the personal lens of the researcher. Adequate training and supervision of the researcher and the awareness and documentation of his or her preconceptions, leanings and personal characteristics, perhaps in a reflective journal (Ortlipp 2008), can help protect against the misrepresentation of participants' perspectives.

14.9 Intellectual Property, Copyright and Publication Ethics

Ethical considerations do not reduce in importance after the completion of the research study and extend into the phase of publication.

A key issue pertains to ensuring ethical decisions regarding authorship and this should be decided at an early stage of the research process to avoid authorship disputes. Many journals have clearly spelt out guidelines for authorship credit as well as for inclusion in the acknowledgments section of the article. Funding sources and conflicts of interest, if any, need to be disclosed.

Researchers must ensure that the measures used in their studies are used with the author's permission, or are available in the public domain as free to use or purchased as required so that there is no copyright infringement.

During the process of writing a manuscript, it is the responsibility of the authors to ensure that there is no plagiarism, intentional or unintentional. There are software programs that can assist in the process of checking for plagiarism. 'Self-plagiarism' or duplicate publications by authors, either in full or part (without providing appropriate reference) is ethically untenable. Training in research ethics for early career researchers must include a focus on plagiarism which is both more easily possible and more likely to be detected in the age of Internet access and technology.

Case reports of counselling or psychotherapy require careful attention so that identifying details are masked before submitting these for publication (Shevell 2004). This is not as simple as it sounds since the revelations of clients in the therapy room are far greater than those in the typical medical case report. The richness of the details is what makes qualitative case studies useful to fellow practitioners and also makes them more vulnerable to breaches of confidentiality. Bond (2004) suggested another strategy to preserve anonymity by combining experiences with several clients into one representative case study. This ethical dilemma of balancing the right to privacy against the right to pedagogical freedom is indeed a challenging and complex one (Levine and Stagno 2001).

A consortium of editors of 12 biomedical journals issued guidelines that prioritized patients' right to privacy and strict procedures for informed consent (International Committee of Medical Journal Editors 1995). However, Levine and Stagno (2001) raised concerns about extrapolating these guidelines to case reports that discuss psychosocial aspects of experience. Some biomedical journals have a policy that the patient must read and approve the manuscript before submission. Taking a different stance, Levine and Stagno (2001) discuss how the process of seeking written informed consent might have a negative impact on vulnerable clients. Indeed, the client's response to this request would be influenced in different ways by the dynamic of the transference. Practical considerations about when to seek consent are addressed with a suggestion that this is ethically appropriate only after peer and editorial review and acceptance of the manuscript for publication.

Ethical obligations continue even after publication of the research findings (Smith 2003). Data must be stored safely for a period of five years (the time frame requirements may vary in different contexts) so that verification is possible if required. Any errors noticed post-publication which would alter the interpretation of research findings, must be addressed in a correction, retraction or erratum, as required.

A position statement outlining standards for responsible research publications can serve as a useful resource for authors (Wager and Kleinert 2011).

How often do ethical transgressions occur in publication of the results of psychotherapy research? The results of an anonymous survey of 257 researchers, primarily from countries in Europe, North America and South America, throw some light on these practices (Braun and Roussos 2012). A proportion of psychotherapy researchers disclosed ethically questionable behaviours; Withholding key aspects of methodology in papers or proposals (11 %), Using another's words or ideas without giving proper credit (10 %), Inappropriately altering or 'cooking' research data (6 %) and Making up research data (2 %). Authorship credit was often given to a person who did not contribute substantively to the manuscript (39 %) but denying authorship credit was infrequently reported (2 %). Although not common, inappropriate alterations or suppression of research reports in response to pressure from a funding source was also present (2 %). The authors discussed these worrying trends in the context of the need for formal training in research integrity and for psychotherapy researchers to be involved in the development of emerging ethical standards.

Journals vary in their policies for reporting ethical review and practices and this is compounded by the lack of a uniform gold standard in ethical procedures as well as varying interpretations of the available guidelines. Chaturvedi and Somashekar (2009) examined ethical reporting practices among studies published in the Indian Journal of Psychiatry in the years 2000, 2003–2007. There were some encouraging trends with informed consent mentioned in 82 % of the studies, a substantial increase from 51 % in the year 2000. There were variations in the way consent was reported; *written* consent was reported in only 40 %, the content of the consent forms in 17 % and the language of the consent form in 3 % of the studies in which consent was mentioned.

Ethics committee approvals were not reported in three-fourth of the studies and this omission was typically among studies which were not conducted at major institutes in the country. Chaturvedi and Somashekar (2009) discuss possible reasons for this lacuna; these include the possible lack of ethics committees or researcher perceptions about the value or necessity of ethics committee review and approval. These patterns have two important implications- (i) Journals need to explicitly state the requirements for reporting informed consent procedures, details of refusal of consent, ethics committee review and other relevant ethical issues in submitted manuscripts and also integrate this into the checklist for manuscript reviewers; (ii) Researchers could seek approval from Independent Ethics Committees who could examine research proposal and provide a documented review for a fee.

14.10 Multi-site and International Collaborative Research

The trend towards collaborative research across multiple sites and/or multiple disciplines has additional implications for the planning, review process and the conduct of ethically informed studies (American Psychological Association Committee on Human Research 2010).

In these collaborative research designs, each principal investigator and research site carries individual and collective responsibility for protecting participants' rights. Communication between investigators across remote locations is essential to ensure the competency of all team members and the integrity and safety of research processes. A written plan should include provisions for safe maintenance, storage and sharing of data across sites, procedures for reporting any adverse or unanticipated events. Ensuring adherence to ethical guidelines requires additional effort and close monitoring across the various research sites and the inclusion of an additional expert consultant could be beneficial.

In multi-site studies, the review process from different Institutional Review Boards (IRB) could result in divergent recommendation related to the research design, procedure or consent documents. This adaptation to local standards may result in delays to the research project and discussions to address emergent concerns with each IRB can be helpful. The documents pertaining to IRB approval at the lead institution may be enclosed with the application at other sites and this could assist in the process of approvals.

Institutional ethical review boards may also have differing interpretations of ethical standards in different countries and this can present additional complications for cross-cultural international projects. Perceptions of sensitivity of the topic or vulnerability of participants are socially influenced, draw from the cultural context and may be viewed differently through a personal, subjective lens (Noland 2012). Jenkins (2013) cites the example of research on sexual orientation among teenagers; while many research studies across various countries rely on the use of retrospective surveys with adults, McDermott (2010) reported that ethical clearance for some US-based studies did not mandate parental consent for surveys exploring sexual orientation directly with teenagers.

Research on psychosocial interventions in developing or resource-poor countries might give rise to a range of ethical questions (Benatar 2002; Weijer et al. 2011). There are questions about the use of both psychotherapy techniques and assessment methodologies developed in industrialized or culturally different nations. Patel et al. (2011) describe their experiences in adapting an Interpersonal Therapy (IPT) module for use in a mixed rural-urban primary care facility in India. This process involved a multiple stakeholders and mixed methods and resulted in various changes; for example, the use of pictorial material, contextually appropriate methods and metaphors, reducing the focus on certain unacceptable components of IPT and inclusion of family members.

Informed consent procedures and the measurement of outcome would need to be modified for use with illiterate participants. The research design must account for the cultural, linguistic, economic and other disparities between the researchers and the participants. This is not true only for international research projects conducted in developing countries; such barriers can exist between researchers and participants from the same country.

Benatar (2002) discusses the danger of imposing the dominant world view and compromising the autonomy and rights of participants in the context of international research projects. Guidelines for ethically informed research in developing countries must reflect the contextual realities, an understanding of the research agendas of industrialized nations, the complexities of 'informed consent', the need to protect against exploitation and the larger issues of social justice.

14.11 Research Ethics: Information Technology and the Internet

The growth in digital communication networks and research on online psychosocial interventions has brought a range of ethical concerns to the forefront. Ethical review would include careful attention to the suitability and safety of online psychosocial interventions for the participants as well as the training and competence of the counsellors or therapists in delivering web-based interventions. Participants with mental health difficulties might reveal suicidal intent and research designs would be evaluated with respect to the possibility of such high-risk situations and the adequacy of response mechanisms (Childress 2000).

Sieber (2001) cautions about safety issues related to storage of electronic data and vulnerability to hacking. Ethical concerns about confidentiality warrant careful attention to secure systems, firewalls and backup data storage systems (Childress 2000).

The online space makes it difficult to be certain about the identity of research participants; children or adolescents with Internet access should not participate in research for adults. The anonymity of responses can be uncertain since the participants' IP address can be captured.

The growing scope of e-interventions has presented contemporary ethical questions which may be closely connected with legal frameworks which vary across geographies and contexts.

14.12 Institutional Ethical Review Boards: A Critical Role

The institutional ethical review system is designed to provide an independent, objective review of research involving human subjects so that the privilege of conducting human subject research may be maintained. Benatar (2002) outlines three primary roles of review boards. (i) The first function is to assess research proposals to assess the balance of risk and benefit to participants, potential conflicts of interests, adequacy of information about the study and its processes and the protection of client autonomy. Specific attention is paid to the consent procedures and freedom for participants to withdraw from the study without negative consequences; (ii) An important role includes the educating researchers and the community in research ethics using various methods (iii) The monitoring and auditing of research projects includes appropriate responses to ethical misconduct in the interests of public accountability.

Boards are comprised of members from varied professional backgrounds without any conflicts of interest or affiliations with the research proposals submitted for review. Members of institutional ethical review boards must receive adequate training in the process of ethical reviews and remain updated on complex and contemporary issues as well as intersections with changing legal frameworks. One concern is that not all members would be familiar with unique challenges arising from research designs involving psychosocial interventions. Review boards need to have a wider perspective beyond the clinical trials framework and have adequate competence to evaluate ethical issues in qualitative research designs as well.

Review boards have the power to reject research proposals on ethical grounds or to suggest substantial changes in the research design. The ethical review process should happen after a scientific review of the research project but the IRB still has the locus standi to critically review aspects of research design when it concerns ethical issues. Another focus of evaluation is the training and experience of the researcher who is delivering the psychosocial intervention to participants in the study.

The primary role of review boards has also been questioned; "Do IRBs serve to protect the institution rather than protecting potentially vulnerable research participants?" Halse and Honey (2005) described their efforts to obtain ethical approval for research with teenage girls with anorexia without specifying the need for prior parental approval. However, "all ethics committees were immovable on these issues. The concern was not with the legitimacy of the arguments presented but with the financial threat of a litigious parent or care-giver". Noland (2012) discussed the fallout when the participants are not vulnerable but the research topic is viewed as sensitive. He asks if the presence of a sensitive or taboo topic should be enough to categorise research participants as high risk and states that some IRBs function under that assumption. Overall, IRBs must guard against being too risk averse while protecting the rights of human participants.

The review procedures must be streamlined along with a provision of expedited review for minor changes in submitted study protocols and mechanisms for reporting any protocol deviations or adverse events during the study. The implementation of processes to monitor ongoing research projects is one of the challenges faced by IRBs. All review decisions must be documented and dated. It is typically mandatory for both funded and non-funded research involving human participants to receive ethical clearance. There is growing need for more institutions to constitute IRBs and for independent Ethics Committees which can function as a key resource for members of the professional community who are not affiliated with an organization.

Institutional Ethical Review Boards increase the accountability of researchers and provide support in our goal to ensure that the scientific merit of research is matched by its ethical merit (Benatar 2002).

14.13 Strategies to Address Ethical Issues in Research

Despite awareness of ethical guidelines and approval from an Institutional Ethical Review Board, ethical dilemmas may arise during the process of research. Haman and Hollon (2009) discuss a range of ethical questions encountered during a large psychotherapy study and listed a range of strategies to anticipate and address these emergent issues.

(i) *Fall back on general ethical principles* The core principles of Beneficence and Nonmaleficence, Fidelity and Responsibility and Respect for People's Rights and Dignity can serve to guide researchers when faced with shifting questions. Dialogic involvement of relevant stakeholders, reflective awareness and acceptance of responsibility are key for researchers in the face of dilemmas.
(ii) *Problem solve in advance* Careful advance planning, clear cut guidelines and the appointment of a neutral external advocate are potentially useful strategies.
(iii) *Use consultations* Various levels of consultation and supervision are advisable.
(iv) *Document* The documentation of difficult decisions, actions with the accompanying rationale is critical.
(v) *Remember why we do research* Finally, being aware of the potential benefits of research; on care and service delivery and indeed a better understanding of ethical complexities; can help balance and address emergent ethical conflicts.

Acknowledgements The authors express their thanks to Prof. S.K. Chaturvedi (Departments of Psychiatry and Mental Health Education; Dean, Behavioural Sciences and Member Secretary, Institutional Ethics Committee, NIMHANS, Bengaluru), for his helpful comments on an earlier version of this chapter.

References

Agarwal, A. (2010). A review of Indian psychiatry research and ethics. *Indian Journal of Psychiatry., 52*(1), S297–S305.

Appelbaum, P. S., Grisso, T., Frank, E., O'Donnell, S., & Kupfer, D. J. (1999). Competence of depressed patients for consent to research. *The American Journal of Psychiatry, 156*(9), 1380–1384.

American Psychological Association. (2010). *Ethical principles of psychologists and code of conduct*. Washington, DC: Author.

American Psychological Association Committee on Human Research. (2010). *Interdisciplinary and multi-site studies: Ethical challenges for psychologists*. Retrieved September 5, 2015 from http://www.apa.org/science/about/psa/2010/11/ethical-challenges.aspx

Arean, P. A., & Alvidrez, J. (2002). Ethical considerations in psychotherapy effectiveness research: choosing the comparison group. *Ethics and Behavior, 12*(1), 63–73.

Becker-Blease, K. A., & Freyd, J. J. (2006). Research participants telling the truth about their lives: The ethics of asking and not asking about abuse. *American Psychologist, 61*(3), 218–226. doi:10.1037/0003-066X.61.3.218.

Benatar, S. R. (2002). Reflections and recommendations on research ethics in developing countries. *Social Science and Medicine, 54*, 1131–1141.

Binik, A., Weijer, C., McRae, A. D., Grimshaw, J. M., Boruch, R., Brehaut, J.C., Allan Donner, A., Eccles, M. P., Saginur, R., Taljaard, M., & Zwarenstein, M. (2011). Does clinical equipoise apply to cluster randomized trials in health research? *Trials*. Retrieved October 27, 2015 from http://www.trialsjournal.com/content/12/1/118

Bolton, P., Bass, J., Neugebauer, R., Verdeli, H., Clougherty, K. F., Wickramaratne, P., et al. (2003). Group interpersonal psychotherapy for depression in rural Uganda. *Journal of American Medical Association, 289*(23), 3117–3124.

Bond, T. (2004). Ethical guidelines for researching counselling and psychotherapy. *Counselling & Psychotherapy Research, 4*(2), 10–18.

Bourdeau, B. (2000). Dual relationships in qualitative research. *The Qualitative Report, 4*(3&4). Retrieved February 26, 2016 from http://www.nova.edu/ssss/QR/QR4–3/bourdeau.html

Braun, M., & Roussos, A. J. (2012). Psychotherapy researchers: reported misbehaviors and opinions. *Journal of Empirical Research on Human Research Ethics, 7*(5), 25–29.

Chaturvedi, S. K. (2008). A review of Indian publications on ethical issues regarding capacity, informed consent, and placebo controlled trials. *Internet Journal of Mental Health, 5*(2). Retrieved September 10, 2015 from http://print.ispub.com/api/0/ispub-article/7089

Chaturvedi, S. K., & Somashekar, B. S. (2009). Reporting ethical aspects in published research articles in the Indian Journal of Psychiatry. *Indian Journal of Psychiatry, 51*, 34–37.

Chauhan, R. C., Purty, A. J., & Singh, N. (2015). Consent for audio-video recording of informed consent process in rural South India. *Perspectives in Clinical Research, 6*(3), 159–162.

Childress, C. A. (2000). Ethical issues in providing online psychotherapeutic interventions. *Journal of Medical Internet Research, 2*(1), e5.

Cwikel, J., & Hoban, E. (2005). Contentious issues in research on trafficked women working in the sex industry: Study design, ethics, and methodology. *The Journal of Sex Research, 42*(4), 306–316.

Etherington, K. (1996). The counsellor as researcher: Boundary issues and critical dilemmas. *British Journal of Guidance & Counselling, 24*(3), 339–346. doi:10.1080/03069889600760311.

Fairburn, C. G., & Cooper, Z. (2011). Therapist competence, therapy quality, and therapist training. *Behavior Research and Therapy, 49*(6–7), 373–378.

Ford, N., Mills, E. J., Zachariah, R., & Upshur, R. (2009). Ethics of conducting research in conflict settings. *Conflict and Law*, 3, 7. Retrieved September 5, 2015 from http://www.conflictandhealth.com/content/3/1/7

Freedland, K. E., Mohr, D. C., Davidson, K. W., & Schwartz, J. E. (2011). Usual and unusual care: existing practice control groups in randomized controlled trials of behavioral interventions. *Psychosomatic Medicine, 73*(4), 323–335.

Freedman, B. (1987). Equipoise and the ethics of clinical research. *New England Journal of Medicine, 317*(3), 141–145.

Fulford, K. W., & Howse, K. (1993). Ethics of research with psychiatric patients: principles, problems and the primary responsibilities of researchers. *Journal of Medical Ethics, 19*(2), 85–91.

Garfield, S. L. (1987). Ethical issues in research on psychotherapy. *Counseling and Values, 31*(2), 115–125.

Grafanaki, S. (1996). How research can change the researcher: The need for sensitivity, flexibility and ethical boundaries in conducting qualitative research in counselling/psychotherapy. *British Journal of Guidance & Counselling, 24*(3), 329–338. doi:10.1080/03069889600760301.

Gandhi, Z., & Vankar, G. K. (2007). Treatment related decision making capacity in psychotic inpatients. *Indian Journal of Psychiatry, 49*, 66–67.

Grisso, T., & Appelbaum, P. S. (1998). *Assessing competence to consent to treatment: A guide for physicians and other health professionals*. New York: Oxford University Press.

Gupta, U. C., & Kharawala, S. (2012). Informed consent in psychiatry clinical research: A conceptual review of issues, challenges, and recommendations. *Perspectives in Clinical Research, 3*(1), 8–15.

Halse, C., & Honey, A. (2005). Unraveling ethics: Illuminating the moral dilemmas of research ethics. *Signs, 30*(4), 2141–2162. doi:10.1086/428419.

Haman, K. L., & Hollon, S. D. (2009). Ethical considerations for cognitive-behavioral therapists in psychotherapy research trials. *Cognitive and Behavioral Practice, 6*(2), 153–163.

Hart, N., & Crawford-Wright, A. (1999). Research as therapy, therapy as research: Ethical dilemmas in new-paradigm research. *British Journal of Guidance & Counselling, 27*(2), 205–214. doi:10.1080/03069889908256265.

Hindmarch, T., Hotopf, M., & Owen, G. S. (2013). Depression and decision-making capacity for treatment or research: a systematic review. *BMC Medical Ethics*, *14*(1). doi:10.1186/1472-6939-14-54

Indian Association of Clinical Psychologists. (2015). *Ethics and code of conduct of clinical psychologists Guidelines 2012–13*. Retrieved June 19, 2015 from http://www.iacp.in/node/159.

International Committee of Medical Journal Editors. (1995). Protection of patients' right to privacy. *British Medical Journal, 311*, 1272

Indian Council of Medical Research. (2006). Ethical guidelines for biomedical research on human participants. Retrieved October 26, 2015 from icmr.nic.in/ethical_guidelines.pdf

Jenkins, P. (2013). The only way is ethics? University ethics committees and the future of psychotherapy research. *Contemporary Psychotherapy*. Retrieved on Septemeber 7, 2015 from http://www.contemporarypsychotherapy.org/volume-5-no-2-winter-2013/the-only-way-is-ethics/

Kass, N. E., Faden, R. R., Goodman, S. N., Pronovost, P., Tunis, S., & Beauchamp, T. L. (2013). The research-treatment distinction: a problematic approach for determining which activities should have ethical oversight. *Hastings Center Report, 43*(1), 4–15.

Levine, S. B., & Stagno, S. J. (2001). Informed consent for case reports the ethical dilemma of right to privacy versus pedagogical freedom. *Journal of Psychotherapy Practice and Research, 10*(3), 193–201.

Lindsey, R. T. (1984). Informed consent and deception in psychotherapy research: An ethical analysis. *The Counseling Psychologist, 12*(3), 79–86.

McDermott, E. (2010). *Researching and monitoring adolescent sexual orientation*. London: Equalities and Human Rights Commission.

McLeod, J. (2001). *Qualitative research in counselling and psychotherapy*. London: Sage.

Muthuswamy, V. (2005). Ethical issues in HIV/AIDS research. *Indian Journal of Medical Research, 121*(4), 601–610.

Noland, C. M. (2012). Institutional barriers to research on sensitive topics: Case of sex communication research among university students. *Journal of Research Practice*, *8*(1), Article-M2.

O'Leary, K. D., & Borkovec, T. D. (1978). Conceptual, methodological, and ethical problems of placebo groups in psychotherapy research. *American Psychologist, 33*(9), 821–830.

Ortlipp, M. (2008). Keeping and using reflective journals in the qualitative research process. *The Qualitative Report., 13*(4), 695–705.

Papas, R. K., Gakinya, B. N., Baliddawa, J. B., Martino, S., Bryant, K. J., Meslin, E. M., & Sidle, J. E. (2012). Ethical issues in a stage 1 cognitive-behavioral therapy feasibility study and trial to reduce alcohol use among HIV-infected outpatients in Western Kenya. *Journal of Empirical Research on Human Research Ethics, 7*(3), 29–37. doi:10.1525/jer.2012.7.3.29.

Patel, V., Chowdhury, N., Rahman, A., & Verdeli, H. (2011). Improving access to psychological treatments: Lessons from developing countries. *Behaviour Research and Therapy, 49*(9), 523–528.

Permuth-Way, J., & Borenstein, A. R. (2009). Financial remuneration for clinical and behavioral research participation: Ethical and practical considerations. *Annals of Epidemiology, 19*(4), 280–285.

Richards, H. M., & Schwartz, L. J. (2002). Ethics of qualitative research: are there special issues for health services research? *Family Practice, 19*(2), 135–139.

Shevell, M. I. (2004). The ethics of case reports. *Paediatric and Child Health, 9*(2), 83–84.

Sieber, J. E. (2001). *Summary of human subjects protection issues related to large sample surveys*. U.S. Department of Justice, Bureau of Justice Statistics. Available from http://www.bjs.gov/content/pub/pdf/shspirls.pdf

Smith, D. (2003). Five principles for research ethics: Cover your bases with these ethical strategies. *APA Monitor on Psychology*, *34*(1), 56. Retrieved September 10, 2015 from http://www.apa.org/monitor/jan03/principles.aspx

Waligora, M., Dranseika, V., & Piasecki, J. (2014). Child's assent in research: Age threshold or personalisation? *BMC Medical Ethics, 15*, 44. doi:10.1186/1472-6939-15-44.

Wager, E. & Kleinert, S. (2011). Responsible research publication: International standards for authors. In M. Tony & S. Nicholar (Eds.), *Promoting research integrity in a global environment* (pp. 309–316). Singapore: World Scientific Publishing Co. Lts. Available from http://www.oeawi.at/downloads/International%20standards_authors_for%20website_11_Nov_2011.pdf

Weijer, C., Grimshaw, J. M., Taljaard, M., Binik, A., Boruch, R., Brehaut, J. C., et al. (2011). Ethical issues posed by cluster randomized trials in health research. *Trials, 12*, 100. doi:10.1186/1745-6215-12-100.

Weijer, C., Grimshaw, J. M., Eccles, M. P., McRae, A. D., White, A., Brehaut, J. C. & Taljaard, M. (2012). The ottawa statement on the ethical design and conduct of cluster randomized trials. PLOS Medicine. Retrieved October 27, 2015 from http://journals.plos.org/plosmedicine/article?id=10.1371/journal.pmed.1001346

Wendler, D., Rackoff, J. E., Emanuel, E. J., & Grady, C. (2002). The ethics of paying for children's participation in research. *Journal of Pediatrics, 141*(2), 166–171.

West, W. (2002). Some ethical dilemmas in counselling and counselling research. *British Journal of Guidance & Counselling, 30*(3), 261–268. doi:10.1080/0306988021000002308.

Wig, N. N. (2004). Ethical issues in psychiatry. *Indian Journal of Medical Ethics, 1*(3), 83–84.

Wiles, R. (2013). *What are qualitative research ethics?*. NY: Bloomsbury.

Wing, J. (1981). Ethics and psychiatric research. In S. Bloch & P. Chodoff (Eds.), *Psychiatric ethics*. New York: Oxford University Press.

World Medical Association. (2001). World Medical Association Declaration of Helsinki. Ethical principles for medical research involving human subjects. *Bulletin of the World Health Organization, 79*(4), 373–374.

World Medical Association. (2013). Declaration of Helsinki: Ethical principles for medical research involving human subjects. *Journal of American Medical Association, 310*(20), 2191–194.

Appendix A

A.1 Professional Codes of Ethics

Mental health professionals encounter ethical dilemmas during their therapeutic practice across the different stages of their professional growth and development. Many professional groups and associations have developed ethical codes of conduct, representing high standards of 'good practice' to help practitioners navigate through ethical dilemmas. Some codes are specific to professional groups (e.g. clinical psychology, psychiatric social work, psychiatry, counselling) while others are related to a therapy modality (e.g. marital and family therapy) or a therapy platform (e.g. online therapy). The codes vary in their coverage and the degree of accountability for any ethical violations by a mental health practitioner. Ethics codes speak to various aspects of professional functioning including practice, research and training but much of it applies to therapeutic encounters with clients or patients. These codes are often revised at varying intervals to incorporate recent developments in the field.

Therapy practitioners should refer to ethical frameworks of the codes for guidance on how to identify and deal with ethical issues and dilemmas and keep themselves updated about any revisions and modifications. The following is a list of some commonly referred ethical codes, along with information on how to access them.

A.2 International Codes of Ethics

A.2.1 American Psychological Association (APA) Ethical Principles of Psychologists and Code of Conduct

The APA Ethics code for psychologists was first published in 1953 and has undergone ten revisions till the current 2010 version of the code. The code was developed based on a study undertaken by APA in 1940, and later replicated in 1992 (Pope and Vetter 1992), asking APA members to describe critical incidents

P. Bhola and A. Raguram (eds.), *Ethical Issues in Counselling and Psychotherapy Practice*, DOI 10.1007/978-981-10-1808-4

that they found to be ethically challenging. The code consists of an Introduction, a preamble, five general principles; beneficence and non-maleficence, fidelity and responsibility, integrity, justice and respect for people's rights and dignity; and specific ethical standards. The principles help uphold the highest professional values and ethical standards in the practice of clinical psychology. The ethical standards tap specific areas such as resolution of ethical issues, competence, human relations, privacy and confidentiality, advertising, sexual relationship with clients and supervisees, research and therapy. A section also addresses the appropriate responses to a colleague's unethical conduct. The APA has recently added information to cover recent developments in the field like telepsychology.

The ethics code can be accessed on the APA website (www.apa.org) and currently at https://www.apa.org/ethics/code/principles.pdf.

Information related to telepsychology can be accessed at http://www.apapracticecentral.org/ce/guidelines/telepsychology-guidelines.pdf.

A.2.2 British Association for Counselling and Psychotherapy (BACP) Ethical Framework for Good Practice in Counselling and Psychotherapy

The BACP Code of Ethics was developed in 2001 and revised in 2002, 2007, 2009, 2010 and 2015. The code stresses upon being mindful and accountable for ethical practice. The code starts with the commitment to client section, followed by two sections—ethics for psychotherapy and counselling and guidance on good practice in counselling and psychotherapy and a separate section on professional conduct procedure. The first section draws attention to ethical diversity, role of culture and setting and enlists a set of values, principles and personal moral qualities in the context of ethical decision-making. The second section covers standards of good practice including competence, trust and confidentiality. It also includes important areas such as working with colleagues and need for personal self-care and probity in professional practice. It primarily highlights principles of justice, extending to being mindful about personal bias and prejudice affecting professional relationship with clients and colleagues. The code also speaks about contemporary issues like homosexuality and opposes any mode of treatment, e.g. 'reparative treatment', which views homosexuality as a mental disorder. The section on professional conduct procedure describes the procedures for submitting complaints about ethical violations and seeking resolution.

The ethics code can be accessed on the BACP website (http://www.bacp.co.uk/) and currently at http://www.bacp.co.uk/admin/structure/files/pdf/14489_ethical-framework-jun15-mono.pdf.

A.2.3 *National Association of Social Workers (NASW) Code of Ethics*

The NASW Code of Ethics is an ethical guideline for social workers accepted worldwide. It was first published in October 1960 in the U.S.A. and underwent multiple revisions, the last revision being in 2008. The code has four sections: the preamble, highlighting the professional values of psychiatric social workers promoting social justice and social change for the clients; the purpose of the NASW Code of Ethics which explains the relevance and function of the code; a description of ethical principles and ethical standards. The code talks about ethical principles of helping people in need and resolving social problems, advocating social justice, respecting dignity of a person, recognizing the importance of human relationships and working with clients in a trustworthy and competent manner. Each principle is tied to a particular core value representing the profession. The ethical standards spell out the psychiatric social worker's ethical responsibility to the client, colleague and practice setting. The section on informed consent has a specific emphasis on steps to be taken with clients who are illiterate or have difficulty understanding the primary language used in the practice setting. The importance of cultural competence and sensitivity is highlighted. The use of any kind of physical touch during the therapy session is specifically prohibited. There is a detailed section on confidentiality pertaining to any media interface and also mentions that client consent is required even for discussions with a consultant. The need to address or report any unethical conduct of colleagues is emphasized.

It can be accessed on the NASW website (http://www.naswdc.org/) and currently at http://www.naswdc.org/pubs/code/default.asp.

A.2.4 *The American Psychiatric Association (APA) Code of Ethics*

The APA code of ethics was developed in 1973 and revised several times till 2013. The code has a preamble and nine sections. It talks about the basic ethical principles of autonomy, justice and beneficence as the bedrock of the client–psychiatrist relationship. Much like other codes, it addresses issues concerning competence and outlines the need to refer or seek supervision when faced with uncertainty or inability to work with the client adequately. It highlights the 'dual roles' of psychiatrists; defined as the competing roles played as a practitioner and as part of an institution. Psychiatrists are obligated to inform patients about the potential for competing obligations either within the treatment or other non-clinical evaluation, such as a forensic evaluation. The code highlights involuntary psychiatric treatment in the larger public interest and to ensure care and protection of the patient. The potential ethical dilemma between autonomy versus beneficence and non-maleficence in these scenarios requires a thoughtful and sensitive approach. It is recommended that any unethical conduct by a colleague be addressed by either seeking supervision, discussing with the colleague or reporting to the concerned regulatory authority.

The ethics code can be accessed on the APA website (http://www.psychiatry.org/) and currently at http://www.psychiatry.org/psychiatrists/practice/ethics.

A.2.5 American Counselling Association (ACA) Code of Ethics

The ACA code of ethics was developed in 1961 and revised several times till the latest revision of 2014. The code begins with a preamble and purpose, and is then divided in eight sections covering the counselling relationship, confidentiality, professional responsibility, relationship with other professionals, assessment, training, research and publication, and distance counselling technology and social media. It has a separate section on resolution of ethical issues, followed by a glossary of terms. The code is unique in its permission to allow dual relationships in select situations when it is beneficial to the client. It also has a section addressing the needs of the terminally ill and in end-of-life care treatment as well as protection of the confidentiality of the deceased. An effort has been made to make the code more culture-sensitive. It further states that when clients have difficulty understanding the language used by the therapist, they must be provided necessary services (e.g. arranging for a qualified interpreter or translator) to ensure adequate comprehension. The revised code permits counsellors to refrain from making and recording a formal diagnosis if they believe that reporting might cause harm to the clients or others. The last revision of 2014 has added an entire section on ethical issues related to online interaction with clients which is detailed and informative.

It can be accessed on the ACA website (http://www.counseling.org/) and currently at http://www.counseling.org/Resources/aca-code-of-ethics.pdf.

A.2.6 American Association of Marriage and Family Therapy (AAMFT) Code of Ethics

The AAMFT code of ethics was first adopted in 1962 and underwent multiple revisions till the last revision of 2015. The code, at the very onset, mentions 'both law and ethics govern the practice of marriage and family therapy' in the preamble. It then enlists a set of aspirational core values. The next section focuses on ethical standards which are defined as rules and standards that a marriage or family therapist is obliged to abide by. These include the following: responsibility to clients, confidentiality, competence and integrity, responsibility to students, research participants and the profession, financial arrangements and advertising. Each section is exceptionally detailed. Confidentiality issues are unique to this therapy modality as the client in question may be more than one, and it is the responsibility of the therapist to respect confidentiality of each of the clients. Competence also involves knowledge about the relevant laws. There is a mention of ethical standards regarding accepting gifts where therapists are expected to be sensitive about cultural norms while deciding whether to accept gifts from clients or not. The latest revision of 2015 has included the ethical standards of technology assisted services. This is a comprehensive section covering requirements of ethical practice including consent, confidentiality, law and location of practice, and training for practising e-therapy. A separate section on Professional Evaluations highlights the need for written informed consent from the clients about the evaluation process, competence in the context of providing testimony in legal systems, the need for accuracy and objectivity while providing expert witness and the need to avoid any conflict between therapy and evaluation, including custody evaluation.

It can be accessed on the AAMFT website (http://www.aamft.org/) and currently at https://www.aamft.org/imis15/Documents/Legal%20Ethics/Board%20Approved%20Code%20for%20Web%20Secured.pdf.

A.2.7 *Employee Assistance Professionals Association (EAPA) Code of Ethics*

The EAPA code developed in 2009 seeks to foster the highest ethical standards among professionals involved in Employee Assistance Programmes (EAP). The employee assistance work is unique from an ethical standpoint considering its 'dual client' nature where both the work organization as well as the individual employee are defined as the 'client'. The code describes the ethical principles of service, beneficence, fidelity, integrity, respect for client's rights and competence. The code talks about professional competence including adequate awareness of the employer organization, human resources management, EAP policy and administration, and EAP direct services. Confidentiality of clients should be maintained where 'clients' include individual employees as well as the employer company or organization. The code is then divided into separate sections covering responsibility to colleagues and other professionals, to employees as clients, to employers or work organizations, to vendors and providers, responsibility as professionals in conducting research and to the employee assistance profession and broader society. The EAPA has a separate section on business practices relevant to this work domain, which discusses aspects of the relationships with suppliers and vendors. The association has also developed an ethical decision-making model that is unique to the needs of this domain of work. An expert Ethics Panel can be contacted for consultation on any ethical quandaries encountered by the members of the association.

The code can be accessed on the EAPA website (http://www.eapassn.org/) and currently at http://www.eapassn.org/Portals/11/Docs/About/EAPACodeofEthics0809.pdf.

A.2.8 International Society for Mental Health Online (ISMHO) Principles for Ethical Online Practice

A joint committee of the International Society for Mental Health Online (ISMHO) and the Psychiatric Society for Informatics (PSI) was formed to develop ethical principles to address the needs of the emerging field of e-therapy. It was endorsed by ISMHO on 9 January 2000 and by PSI on 13 May 2000. The principles involve informed consent (about the process, the clinician, the potential risks and benefits, safeguards and alternatives), standard operating procedure (competence, legal requirements, the structure of the services, evaluation, multiple treatment providers, confidentiality, records, and existing guidelines), and emergencies (procedures and local backup). These issues are discussed with special reference to practitioner–client interactions in the online space.

The code of ethics can be accessed on the ISMHO website (http://www.ismho.org/).

A.3 Indian Codes of Ethics

In India, different mental health disciplines have adopted codes of ethics to assist ethical practice among professionals. The psychiatric social work discipline refers to the NASW code of ethics for ethical guidance.

A.3.1 National Academy of Psychology Ethical Principles for Psychologists

The NAOP adopted the ethical principles for psychologists in December 2010. The principles are in accordance with the Universal Declaration of the Ethical Principles for Psychologists by the International Association of Applied Psychology in Berlin in July, 2008. It provides a brief moral or philosophical framework for ethical clinical practice including respect for people, caring for their wellbeing, integrity of the professional and responsibilities to the society. This framework is aimed at sensitizing practising psychologists about the ethical responsibilities and concerns in their professional roles.

It can be accessed on the NAOP website (http://www.naopindia.org/) and currently at http://www.naopindia.org/ethical-principles.

A.3.2 Indian Association of Clinical Psychologists (IACP) Ethics and Code of Conduct of Clinical Psychologists

The IACP code of ethics is an ethical guideline for clinical psychologists upholding the highest standards of professional practice. The initial 1995 IACP code of ethics was brief and sketchy, but this is in the process of revision. A draft version is available on the association website. The revised draft provides a detailed description of the general principles of professional competence, professional responsibility including social responsibility and integrity, confidentiality and avoiding exploitation of clients. The current definition of Competence in the Indian context mandates practising clinical psychologists to be registered with the Rehabilitation Council of India. Confidentiality and specific situations that warrant disclosure, informed consent and dual relationships are discussed. The code addresses the need to be aware of laws and governments statutes and ethical roles and responsibilities in forensic work. It further draws attention to psychological assessment techniques and research as well as ethical concerns related to therapy. The code also has a brief section covering contemporary concerns arising in the e-therapy modality.

It can be accessed on the IACP website (www.iacp.in) and currently at http://www.iacp.in/node/159.

A.3.3 Indian Psychiatric Society (IPS) Code of Ethics

A committee was appointed by the Indian Psychiatric Society to prepare the code of ethics for psychiatrists in India and this was approved at its annual conference in Cuttack in 1989. The code presents a set of principles upholding high moral standards. According to this code, benevolence and interest of the patient must precede self-interest. The code emphasizes the need for informed consent for treatments, except in emergency situations or with family consent for drug abusers who refuse treatment. Most codes do not explicitly address the issue of gifts but the IPS code explicitly prohibits the acceptance of "gifts or gratifications from patients under treatment". The IPS guidelines speak of sexual advances toward patients as unethical. The importance of respecting the human rights of mentally retarded clients is mentioned. A brief outline of the code is available as part of a psychiatry textbook (Ahuja 2006). The ethics code of the Medical Council of India (MCI), freely accessible online, applies to all medical professionals, including psychiatrists.

Ahuja (2006). A short textbook of psychiatry. 6th Ed. New Delhi: Jaypee Brothers Publishers.

The MCI code of ethics can be accessed at http://www.mciindia.org/Rules-and-Regulation/Ethics%20Regulations-2002.pdf.

Appendix B

B.1 Sample Informed Consent Form

The following topics have been discussed adequately with me:

- The nature and process of therapy/counselling, including potential risks and benefits and estimated duration of contact.
- My role and what I can expect from my therapist/counsellor.
- The goals we could work towards.
- Treatment options that are available to me and their pros and cons.
- Details about fees, scheduling and cancelling appointments, and Dr./Mr./Ms. …….'s availability and contact procedure.
- The confidentiality I can expect (with verbal information and therapy records) and limits to the same.
- My right to terminate treatment if I wish to and my right to withdraw consent.

I agree to the terms discussed and further, commit to the following during the course of therapy/counselling:

1. ……………………………………………………………………

   ………………………………….

2. ……………………………………………………………………

   …………………………………...

   …………………………………………

   Name and Signature

This form was developed by Dr. Rathna Isaac and is based on the various codes of conduct as well as on the scientific literature. It is ideal for adult individual psychotherapy. Please note that child/adolescent therapy and couples therapy will require additional areas to be addressed. Any informed consent form will need to be printed on the letterhead of the practitioner. The form can to be modified to add items if you find that certain issues come up frequently in your setting, and therefore feel a need to be more explicit about them. However, deletion of any of the items would not be recommended.

P. Bhola and A. Raguram (eds.), *Ethical Issues in Counselling and Psychotherapy Practice*, DOI 10.1007/978-981-10-1808-4

MIX
Papier aus verantwortungsvollen Quellen
Paper from responsible sources
FSC® C105338
FSC
www.fsc.org

If you have any concerns about our products,
you can contact us on
ProductSafety@springernature.com

In case Publisher is established outside the EU,
the EU authorized representative is:
Springer Nature Customer Service Center GmbH
Europaplatz 3, 69115 Heidelberg, Germany

Printed by Libri Plureos GmbH
in Hamburg, Germany